D1500267

New Drug Discovery
and Development

THE WILEY BICENTENNIAL—KNOWLEDGE FOR GENERATIONS

*E*ach generation has its unique needs and aspirations. When Charles Wiley first opened his small printing shop in lower Manhattan in 1807, it was a generation of boundless potential searching for an identity. And we were there, helping to define a new American literary tradition. Over half a century later, in the midst of the Second Industrial Revolution, it was a generation focused on building the future. Once again, we were there, supplying the critical scientific, technical, and engineering knowledge that helped frame the world. Throughout the 20th Century, and into the new millennium, nations began to reach out beyond their own borders and a new international community was born. Wiley was there, expanding its operations around the world to enable a global exchange of ideas, opinions, and know-how.

For 200 years, Wiley has been an integral part of each generation's journey, enabling the flow of information and understanding necessary to meet their needs and fulfill their aspirations. Today, bold new technologies are changing the way we live and learn. Wiley will be there, providing you the must-have knowledge you need to imagine new worlds, new possibilities, and new opportunities.

Generations come and go, but you can always count on Wiley to provide you the knowledge you need, when and where you need it!

WILLIAM J. PESCE
PRESIDENT AND CHIEF EXECUTIVE OFFICER

PETER BOOTH WILEY
CHAIRMAN OF THE BOARD

New Drug Discovery and Development

Daniel Lednicer
North Bethesda, MD

WILEY-INTERSCIENCE
A JOHN WILEY & SONS, INC., PUBLICATION

Published by John Wiley & Sons, Inc., Hoboken, New Jersey
Published simultaneously in Canada

For general information on our other products and services or for technical support, please contact our Customer Care Department within the United States at (800) 762-2974, outside the United States at (317) 572-3993 or fax (317) 572-4002.

Wiley also publishes its books in a variety of electronic formats. Some content that appears in print may not be available in electronic formats. For more information about Wiley products, visit our web site at www.wiley.com.

Library of Congress Cataloging-in-Publication Data is available

ISBN-13 978-0-470-00750-1
ISBN-10 0-470-00750-8

Printed in the United States of America

10 9 8 7 6 5 4 3 2 1

To the human curiosity that long ago led some individual to taste those berries and then millenia later prompted a chemist to follow a hunch in designing a molecule.

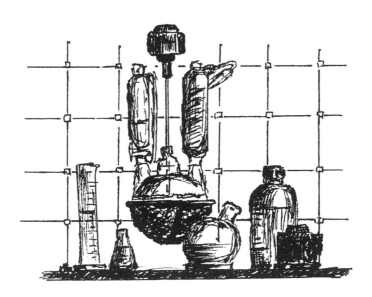

Contents

Preface

Therapeutic agents, more often called drugs, play an increasingly important role in our aging population. This is reflected most directly in the ever larger portion of family income spent on medicines, an increase that is only partly attributable to the rising cost of filling a prescription. The public and much of the media point to the greed of "big pharma," the small number of huge pharmaceutical companies, as the main cause of this price escalation. The companies in turn defend prices by citing the cost of research aimed at finding and developing new therapeutic agents. It can thus be instructive to set aside the heat of the argument and to take a closer look at the historical origins of some representative examples of the extensive pharmacopoeia available to physicians for treating their patients.

To that end, this volume presents a set of case histories that have led to families of drugs for treating many of mankind's ills. Some, for example the antibiotics, have contributed to a marked increase in longevity in the advanced world over the past half century. The effect of some other categories, for example the central analgesics, has been on patients' quality of life. The discovery of a new drug category inevitably leads to the development of related compounds as companies race to enter the burgeoning market. This account notes some of the entries that came after the pioneering drug, although not in the detail found in more specialized volumes.

The first chapter traces the development of the major classes of antibiotics used to treat infectious disease. The case histories provide interesting examples of the interplay between guided research and serendipity that marks much of drug research. The synthesis of the first effective antibacterial agent, the dye Prontosil, was prompted by the well-known affinity of stains for bacteria. It was found not much later that activity was in fact due to colorless sulfanilamide produced by cleavage of the dye by liver enzymes. The discovery and early development of penicillin took place in university laboratories. The antibiotic became a usable drug through involvement of industry at later stages. The second chapter deals with agents for treating the other major aspect of infectious disease, the virus. The great majority of antiviral drugs are of much more recent origin than antibiotics. This field of research interestingly received major stimulus from the AIDS epidemic. Some of the first antiviral drugs had their origin in anticancer research programs and reflect the reliance on screens that tested a wide variety of chemicals characteristic of the programs. The more recent antiviral compounds on the other hand were designed to take advantage of newly gained insights into the molecular biology of viral infection.

The somewhat complex development of antihypertensive agents is considered in Chapter 3. Classes of drugs used to treat elevated blood pressure came from two disparate sources. A number were developed on the basis of the knowledge base in physiology and pharmacology that came out of basic research in academia and the NIH as well as industry. This provided the theoretical rationale for the early alpha blockers; the actual chemical compounds were found by random screening of products that came from the labs of organic chemists. In an interesting feedback, new compounds in this class, notably clonidine and prazocin, led to new insights into pharmacology. This motif is repeated in the calcium channel blockers, where a dihydropyridine found by random screening led to the elucidation of a new mechanism of action. The history of the angiotensin-converting enzyme (ACE) inhibitors and the more recent angiotensin antagonists drew heavily on the basic pharmacology of kidney function.

The deleterious effects of high levels of serum cholesterol were at least suspected by the middle of the nineteenth century. The development of compounds designed to lower cholesterol is found in Chapter 4. This seemingly simple end point led to the adoption of relatively straightforward animal screens intended to identify compounds that lowered serum cholesterol. A number of drugs that came out of this program are still in use. Several other agents were carefully designed to take advantage of the oxidation of cholesterol to bile acids by the liver. Acceptance was limited by the fact that treatment comprised taking these unpalatable powders in gram quantities. More recent work has been guided by new information on the biochemical mechanism involved in regulation of blood lipids. This insight led to the discovery of the statins, a class of drugs that has dramatically changed this area of therapy.

The discovery of the pain-killing and soporific activity of the dried exudates of the flower pod of the poppy, papaver somniferum, took place well before recorded history. The isolation of the active principle from opium, as the crude substance was called, dates back to more recent times; that compound, morphine, was known in crystalline form by the early nineteenth century. As organic chemistry developed, chemists turned their attention to this molecule. They sought by manipulating the chemical structure to produce less addicting congeners that were active when taken by mouth. Chapter 5 first traces the development of several classes of analgesics built on the same complex carbon skeleton as the parent natural product. The account then switches to a series of purely synthetic compounds that to an organic chemist bear little resemblance to morphine. Many of these new compounds are several tens of orders of magnitude more potent than morphine. The goal of a nonaddicting opioid has however still not been met.

Nonopiate compounds for treating pain, often called peripheral analgesics, comprise one of the most widely prescribed classes of drugs. This usage is supplemented by consumption of a host of over-the-counter drugs. Drug development in this field has been a largely empirical enterprise. The discovery of the role of prostaglandins in pain and inflammation has guided some of the more recent research.

Drug discovery is anything but a linear process. Synthesis programs carefully designed to produce compounds active in a specific test more than occasionally lead to the discovery of an entirely new class of drugs. The chapter also traces the origin of the COX-2 nonsteroidal anti-inflammatory agents (NSAID) to a much earlier research program on anti-estrogens. The odyssey that weaved its way over the years through many different labs culminated in the discovery of celecoxib, more familiarly known as Celebrex. When first launched, this and its relatives were believed to present a significant advance over previously available NSAIDs.

The drugs whose development is considered in Chapter 7 share a common rather complex chemical structural framework. Small changes in the chemical structures, however, lead to quite disparate activities. The so-called sex steroids comprise a large subdivision of this class, which includes compounds with variously estrogenic, progestational, and androgenic activity. Research on this group was initially motivated by the involvement in reproductive function of steroids that had been isolated from animal blood and tissues. This work, perhaps not surprisingly, culminated in the development of the oral contraceptives. The resulting drugs, known collectively as "The Pill" have arguably brought about enormous changes in societal mores. The androgens and the related anabolic steroids are currently in the spotlight for their putative abuse among professional athletes. The clinical observation of the profound anti-inflammatory activity of high doses of the naturally occurring steroid, cortisone, led to a race among pharmaceutical companies for a practical means for producing the compound in large quantity. Once this was found, a correspondingly large effort was devoted in many labs to finding more potent congeners devoid of the classical side effects of cortisone. The first goal was eventually met, although side effects seemed to be inevitable with activity. The very potent agents were, however, well tolerated when applied topically. Those highly potent steroids are today widely used in anti-allergic nasal sprays.

Therapeutic goals provided the unifying theme for the first half dozen chapters, but the drugs discussed in Chapter 7 share a common structural element, the steroid nucleus. The very diverse set of compounds in Chapter 8 all come about from the search for agents to combat the adverse effects of histamine. This effort led not only to the familiar antihistamines, but also the tricyclic antipsychotic and antidepressant agents as well a series of antiulcer agents. Better understanding of the allergic action led to the recognition of the important role played by leukotrienes. This has resulted in the development of several anti-allergic leukotriene antagonists.

Each of the previous sections glosses over the various activities that take place between the identification of a new drug and its appearance on a pharmacy shelf. This therefore omits the enormous expenditure of time, effort, and money that must be spent in the process. Chapter 9 gives a brief overview of the many steps required before a drug is blessed with approval from the Food and Drug Administration and how these have changed over time.

This book is aimed at the informed layman, that is, an individual who, although not a chemist, has at least some background in the sciences. The use of technical terms and concepts is of course almost unavoidable in a book of this nature.

A deliberate attempt has been made to keep these to a minimum. The author, an organic chemist by training and persuasion, has included structural formulas in this account. The discussion is presented in terms that do not require reliance on those formulas. Structures have been included in the belief that they will provide additional insight to those readers who are conversant with that way of thinking. A brief review of conventions used to depict molecules is included in the Appendix for those whose exposure to organic chemistry was fleeting or largely forgotten.

Dan Lednicer
North Bethesda, Maryland

Chapter 1

Antibiotics

The increase in life expectancy seen during the twentieth century in many parts of the world is by now too familiar to require lengthy discussion. Expectancy at birth in the United States, for example, increased by close to two decades from 49.2 years at the beginning of that century to 68.1 years in 1950.[1] This remarkable jump generally has been attributed to improvements in sanitation and the advent of drugs for the treatment of infectious disease. Many bacterial infections that required hospitalization before World War II are now treated with a course of antibiotics. Usage is so well accepted that treatment will involve prescription called into the pharmacy by the physician's office and self-administration at home. The isolation of pure penicillin in 1939 in England is often used to date the beginning of the development that has led to today's armamentarium of antibiotics drugs. However, the story in fact begins back in Germany in the early 1930s.

It was well recognized by then that the structure and enzyme systems that allow bacteria to thrive are very different from their mammalian counterparts. Scientists devoted considerable effort from the late nineteenth century on to the search of chemicals that would exploit those differences so as to specifically eradicate bacterial cells. A tantalizing early clue that such differences might exist lay in the stains that were used by microbiologists to study their prey. Drawing on the wealth of synthetic dyes produced by the burgeoning chemical industry in Germany in the nineteenth century, scientists had identified a series of substances that stained bacteria in preference to mammalian cells.[2] The key to finding a drug that would preferentially eradicate bacteria seemed to be to find a stain that would kill the cells that accepted that particular dye. The long search seemed to have finally borne fruit in 1932. Gerhard Domagk, working in a laboratory set up by I. G. Farben, discovered that the red dye Prontosil Rubrum (Fig. 1) protected mice that had been injected with otherwise lethal doses of staphylococci.[3] Use of the dye to treat successfully a human infection (Domagk's daughter) confirmed that this was indeed a therapeutic breakthrough. There had by then, however, been more than a few reports of seemingly miraculous cures of disease due to bacterial infection.

New Drug Discovery and Development by Daniel Lednicer
Copyright © 2007 John Wiley & Sons, Inc.

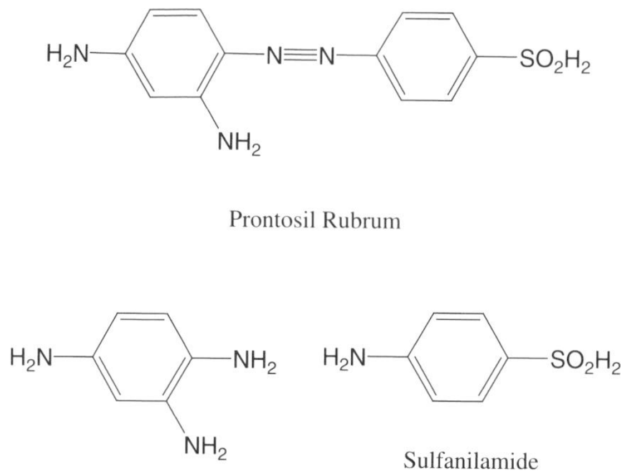

Prontosil Rubrum

Sulfanilamide

Figure 1   Prontosil Rubrum and its conversion to sulfanilamide.

The resulting skepticism from the inevitable failure of those earlier treatments led to surprisingly slow acceptance of the dye, by now simply called Prontosil. There was also the puzzling fact that the dye was only marginally effective in killing bacteria in vitro, that is, in the then standard test tube experiment for antibacterial activity. A group of scientists at the Pasteur Institute in France then showed that the dye molecule is transformed chemically in animals.[4,5] Liver enzymes, they found, split the molecule in two at the central nitrogen to nitrogen azo linkage (N=N). One of the halves, subsequently named sulfanilamide, turned out to be a fully effective antibacterial compound both in test tubes or when administered to infected mice. The other half showed no antibacterial activity whatsoever. This work incidentally gave birth to the discipline of drug metabolism. Prontosil was to be but the first case of a drug that needed to be modified by the body for activity.

An immediate result of this finding was the abandonment of Prontosil in favor of the chemically much simpler sulfanilamide. This drug can be synthesized from benzene in just a few steps.[6] In fact, this synthesis was for many years a laboratory exercise at the beginning of organic chemistry courses. It probably enticed more that one student, including the author, into a career in pharmaceutical research.

The discovery of sulfanilamide marked the beginning of the search for agents to treat infectious disease among compounds made from scratch by organic chemists. We come back to that story later. The discovery of the other important source of compounds that selectively kill microbes dates back to 1929 and Alexander Fleming's well-known serendipitous discovery. He noted a microbe-free clear zone around a mold colony that had contaminated a culture in a Petri dish, and correctly ascribed that to an antibiotic substance secreted by the mold. He named this unknown secretion penicillin after the producing mold, which he had identified as *Penicillium notatum*.[7] The imminence of World War II is said to have spurred the transition of what had been considered as simply an interesting laboratory observation into a useful antibiotic drug. Beginning in about 1938, Howard Florey led the very major effort to isolate penicillin. This was finally accomplished in 1940, largely through the work of his Oxford collaborator Ernst Chain. The team isolated just enough pure penicillin to ascertain its near miraculous activity in humans.[8] Production of penicillin was transferred to the United States, because the British chemical industry was at that time fully tied up with war production. In its original form, *Penicillium notatum* grew best as a surface mat. Production in large quantities invoked visions of the use of shallow tanks with enormous surface areas, so the project was assigned to the U.S. Department of Agriculture Northern Laboratory in Peoria, Illinois, which had experience in industrial fermentation. There they devised a method for growing the mold as a submerged culture. By this and other means they greatly increased the yield of penicillin.[8,9] The method developed by USDA was then transferred to industry, and a large number of companies with expertise and facilities for fermentation were enlisted in the effort.[10] This even at one time included Schenley, better known as a producer of spirits.

Penicillin in the form used then had a number of very serious shortcomings. The drug had to be administered by injection as it was not orally active. The molecule is also very reactive, leading to poor stability. Early research aimed at producing more stable congeners led to several salts with improved stability. Reasons for the sensitivity of penicillins emerged with the determination of their chemical structure. The compounds are in essence comprised of two discrete connected pieces. The essential part consists of a fused ring structure called a beta-lactam. This is the reactive part that in the end kills bacteria; it also contributes to the lack of stability. The rest of the structure, which is also required for activity, consists of an organic acid connected through a chemical bond. Penicillin obtained from fermentation is a mixture of closely related compounds in which the invariant beta-lactam is hooked to slightly different acids. Penicillin G is one of the major components. Scientists had noted that they could increase the proportion of one or another congener by adding a large amount of that acid to the culture medium (Fig. 2).[11,12] This allowed them selectively to produce one or another of the congeners such as Penicillin V. These still, however, shared many of the same shortcomings. This included poor stability and lack of oral activity; the drugs were also not effective against a significant number of classes of bacteria. It had become apparent by 1960 that further improvements would require manipulation of the chemical structure.

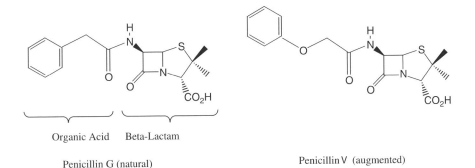

Figure 2   A penicillin from the naturally occurring complex and a typical augmented feeding product.

It was later established that the selectivity of the beta-lactam antibiotics traces back to the fact that bacteria are more closely related to plants than animals. Individual animal cells are surrounded by a membrane, whereas plants and thus bacteria depend on a wall for cell integrity. In bacteria that structure is composed of a dense network of protein filaments that is cross-linked by chemical bonds. A significant number of the amino acids that compose the proteins have chemical structures that are mirror images of those found in animals. The beta-lactams (penicillins and cephalosporins) are mistaken by bacterial enzymes as small pieces that will be used to form the cross-links. Once they get incorporated, they bring the process to a dead halt, causing the cell wall to rupture.[13] The drugs are thus selective, because mammals do not use cell walls and in addition utilize enzymes that do not recognize the mirror image amino acids used to make bacterial cell walls. The beta-lactam enzymes are consequently known for their very large safety margins. There is, however, a distinct portion of the population that is extremely allergic to these drugs.

By 1940 sulfanilamide had come into widespread use particularly in treating war wounds. The drug was used both as a tablet and sprinkled directly onto open lesions. Although the drug saved numerous lives, many types of bacteria were immune to its action. Chemists in a number of pharmaceutical laboratories then tried to make changes in the molecule in attempts to broaden its activity against other classes of bacteria (Fig. 3). Systematic work showed that there was only a single place on the molecule that could be manipulated and still retain antibacterial activity. Hundreds of analogues of sulfanilamide such as sulfathiazole were probably prepared in a number of laboratories between 1940 and the late 1950s, by which time the work was finally abandoned.[14] No fewer than 27 of these were granted nonproprietary names, which is often an indication that the sponsor intends to test the compound in the clinic. At least eight of these so-called sulfa drugs are currently used in the clinic.

The antibacterial activity and selectivity of this class of drugs again depends on the fact that bacteria uniquely rely on biochemical processes that have no counterpart in more complex organisms (Fig. 4). Folic acid, perhaps better known as one of the B vitamins, is an essential factor in various metabolic processes such as formation of red blood cells and DNA itself. Over the course of evolution many organisms have lost the ability to make this compound and rely on obtaining it in food. Bacteria, on the other hand, synthesize this vitamin from scratch. An important biochemical step involves hooking a small molecule, PABA (para-aminobenzoic acid), onto the growing molecule. The chemical structures of the sulfa drug are similar enough to PABA to cause bacterial enzymes to incorporate these molecules.

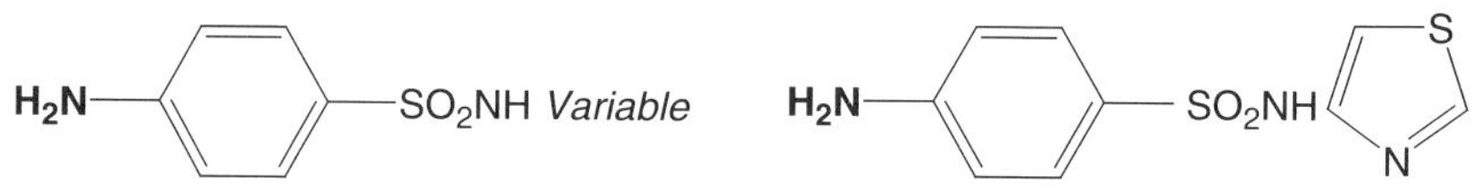

ALLOWED SUBSTITUTION   SULFATHIAZOLE

Figure 3   Sites that may be modified on sulfa drugs.

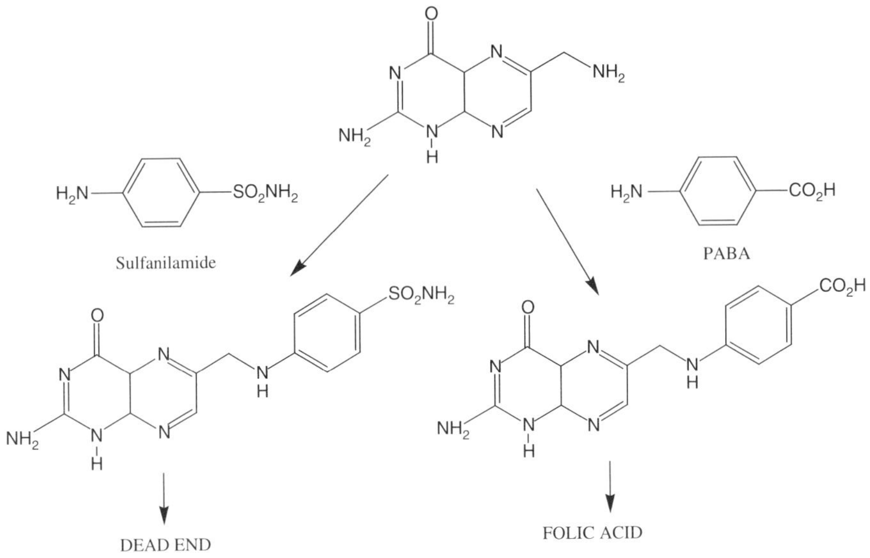

Figure 4 Sulfa drugs' mode of action.

However, the resulting product can go no further, in effect causing the bacterium to die for lack of folic acid.[15]

One of the sulfa drugs, sulfamethoxazole, is a constituent of combination tablets, such as Bactrim®, that still comprise first-line treatment for urinary infections. The other active ingredient, trimethoprim, originates in work carried out by future Nobel Prize winner George Hitchings at Burroughs Wellcome in the late 1950s.[16] Taking their cue from compounds involved in enzyme action, he and his associates prepared a congener called pyrimethamine.[17] This agent proved to inhibit bacterial growth by interfering with an enzyme, dihydrofolate reductase (DHFR), involved further down the line in the synthesis of folic acid. Further work along the same lines led to the synthesis of trimethoprim. The combination tablet exploits the fact that each of the active ingredients inhibits bacterial growth by interfering with different enzymes that bacteria need survive (Fig. 5). This combination has been used for treating HIV/AIDS opportunistic infections.

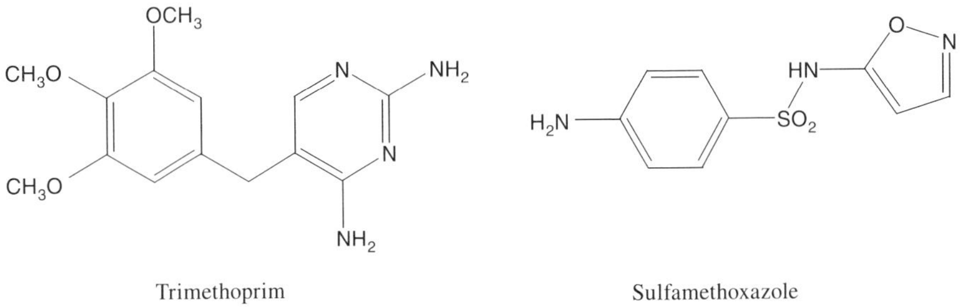

Trimethoprim   Sulfamethoxazole

Figure 5 Active ingredients in Bactrim.

Isolated reports of unusual side effects came with widespread use of sulfa drugs.[18] Very high doses caused some patients to excrete water and others to show a drop in blood sugar levels. Chemists in pharmaceutical laboratories seized on these apparent side effects to develop entirely new classes of drugs. By manipulating the chemical structures, scientists at Hoechst came up with a compound that normalized blood sugar in adult onset diabetics. This drug, chlorpropamide, in which the all-essential amino group is replaced by methyl, is virtually devoid of antibacterial activity. Replacement of the sulfonamide (SO_2NH_2) by a sulfonylurea function ($SO_2NHCONH_2$) proved crucial for antidiabetic activity and was present in the widely marketed drug tolbutamide.[19] The only drugs available in the 1940s for treating conditions that required loss of body water, the diuretics, included mercury in their chemical composition. Scientists at Merck, led by Sprague, were able to introduce changes on the structure of the sulfa drugs to produce a well-tolerated diuretic drug. These principally involved adding an additional sulfonamide onto the parent molecule. This compound, chloraminophenamide, also devoid of antibacterial activity, is no longer in use. It has been superseded by hydrochlorothiazide, which was first synthesized by chemists at Ciba. This drug, often better known by its acronym, HCTZ (Fig. 6), is still used as first-line treatment of patients with mildly elevated blood pressure.[20]

The discovery of penicillin led to the recognition of the ability of fungi to protect themselves against microorganisms by secreting compounds that inhibit bacterial replication and in fact often kill off those threatening organisms. Penicillin itself showed that such antibiotics may act specifically on enzymes that do not have counterparts in mammals. This property, shared with the sulfa drugs, led to low toxicity to humans. The search for new molecules in this class turned to the

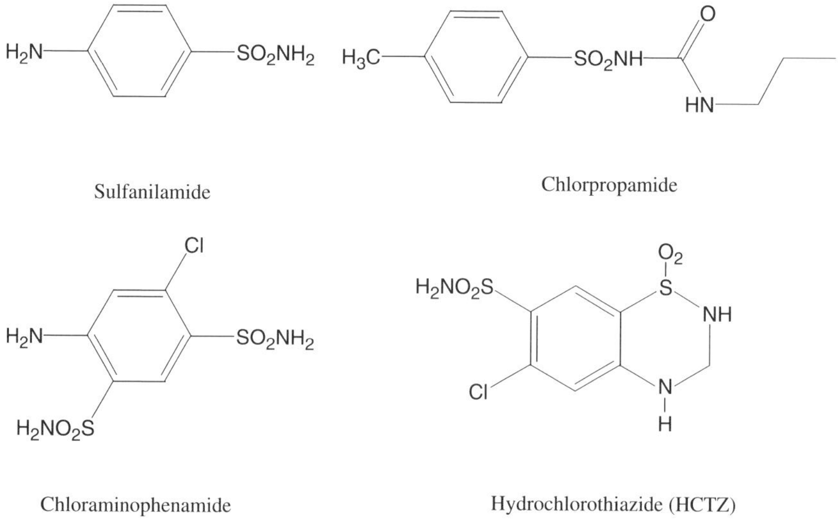

Sulfanilamide

Chlorpropamide

Chloraminophenamide

Hydrochlorothiazide (HCTZ)

Figure 6   Nonantibiotic drugs related to sulfonamides.

investigation of the large number of fungi that inhabit the world. The competition between these organisms and bacteria in soils pointed to that domain as a potentially rich source of antibiotics. In the early 1940s, Albert Schatz, working at Rutgers University under the supervision of his professor, Selman Waksman, discovered that the mold *Streptomyces Griseus*, produced an antibiotic that had a complex sugar-based chemical structure that was quite different from penicillin.[21,22] It was more stable than penicillin and was effective against a somewhat different set of pathogens. This therapeutic agent, streptomycin, is today still one of the indispensable drugs used to combat tuberculosis. This was the first in a series of antibiotics composed of linked sugar-like components. The presence of basic nitrogen sets those fragments apart from simple sugars. These antibiotics, known as aminoglycosides, are largely used for treating severe infections (Fig. 7).

The aminoglycosides are part of a group that acts on the machinery within cells, be they bacterial or mammalian, that assemble the proteins involved in reproduction and growth. To accomplish this task, a central component, called the ribosome, slides along the string of RNA that incorporates the sequential three base codes for amino acids that will from part of the protein chain. Smaller nucleic acids, called transfer RNA, which include only the three-base sequence for individual amino acids, convey those fragments to the ribosome. The aminoglycosides in essence cause the messenger RNA, which carries the template for a new protein, to be misread by the ribosome. This causes the ribosome to cause errors in making new proteins, which are then lethal to the cell. Selective toxicity to bacteria and not the host depends on the species difference in the parts used in this process. Aminoglycosides, like many other fermentation products, often occur as mixtures of very closely related compounds. The commercial form of the antibiotic gentamycin, for example, consists of a mixture of C_1, C_{1A}, and C_2, all of which have similar activity.[23] In order to overcome some of the shortcomings of the aminoglycoside

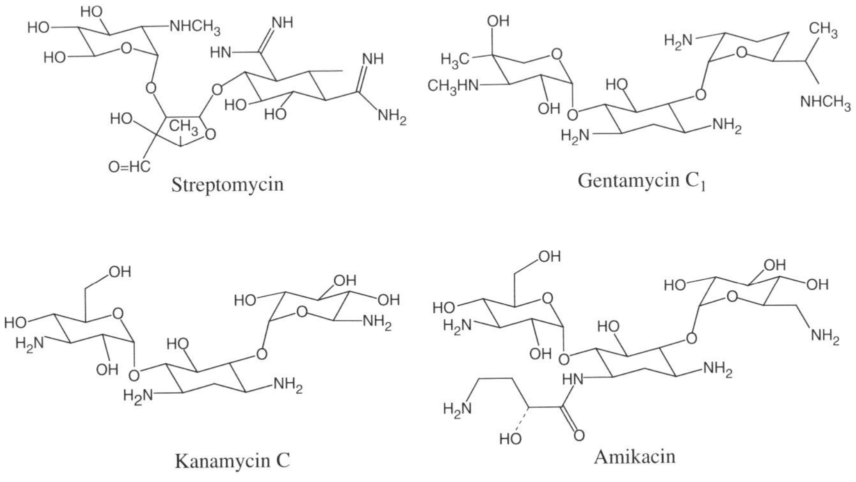

Figure 7   Aminoglycoside antibiotics.

kanamycin, which had first been isolated in 1957, scientists at Bristol–Myers under-took a program to prepare semi-synthetic derivatives. This led to the marketed aminoglycoside amikacin, which differs largely in the incorporation of a new side chain on the central ring.[24] Aminoglycoside antibiotics retain activity against bacteria that have developed resistance to some of the more common drugs because they act on a very different target on the organism.[25] An apocryphal tale has it that the approval of a novel and highly potent aminoglycoside was met with high expectations for commercial success on the part of its sponsor. The very properties of the drug led to an agreement on the part of infectious disease specialists to use the drug extremely sparingly. Setting the antibiotic aside in a reserve category, it was felt, would provide last resort treatment for infections with bacteria with multidrug resistance.

This discovery also launched a major effort in the laboratories of many pharma-ceutical companies to screen soils from a wide variety of sources. Thus, in 1949, Benjamin Duggar of the University of Wisconsin, who was a consultant to the Lederle Laboratories, discovered a new antibiotic that he called aureomycin.[26] Scientists at Lederle used this finding to develop a family of chemically closely related antibiotics. These are called tetracyclines after the chemical structure, which involves four linked rings (Fig. 8). The original drug had a relatively short duration of action due to inactivation in the bloodstream. Chemical manipulation of the natural product led to doxycycline, the first tetracycline that was effective when taken just once per day.[27] This class of drugs also inhibits protein synthesis

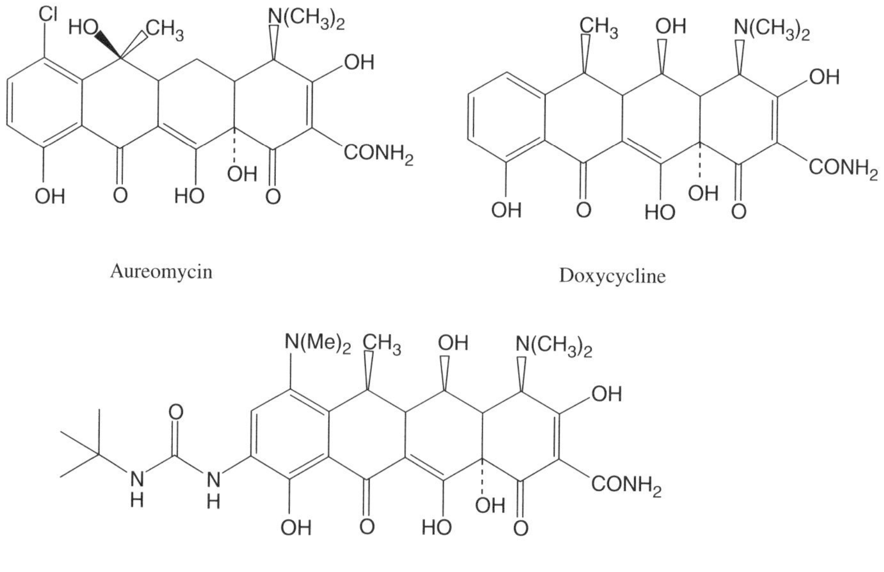

Aureomycin

Doxycycline

Tigecycline

Figure 8   Tetracycline antibiotics.

in bacteria, although at a different biochemical site than do the aminoglycosides. Although the causative agent of malaria is a plasmodium parasite, it is susceptible to doxycycline at one of the stages of its very involved life cycle. The drug thus finds extensive use as a prophylactic for travelers in malaria-infested areas, and is used to treat the disease in combination with other drugs. The tetracyclines exhibited activity against a wide variety of microbes when they were first introduced. Resistant strains, however, developed over time, as has happened with most other classes of drugs. Very recent work in the original laboratories, by the now Wyeth–Ayerst, resulted in a chemically highly modified derivative that is effective even against tetracycline-resistant strains. This compound, tigecycline, was approved for sale in the United States in 2005.[28]

A soil sample from the Phillipines was sent back to the Indianapolis labs of Eli Lilly by one of their local employees at about the same time. Screening of that sample led to the isolation of a new antibiotic with yet another novel chemical structure. The organism, at that time called *Streptomyces erythaeus*, gave the compound its name. (Many Streptomyces species have since been reclassified as Actinomyces for reasons of taxonomic accuracy.) This antibiotic was developed into the still widely used drug erythromycin by a team at Lilly led by J. M. McGuire. The structure is a good bit more complex than preceding antibiotics, and consists of a very large fourteen-membered ring with attached sugars, one of which has a basic amino group (omitted in Fig. 9 for reasons of clarity).[29] This compound too inhibits bacterial protein synthesis, again by a somewhat different mechanism. The relatively short duration of blood levels of the drug encouraged chemists to try to modify the basic structure so as to overcome that property. In the late 1980s a group of chemists at the Croatian (then Yugoslav) pharmaceutical company Sour Pliva succeeded in preparing a derivative in which a basic nitrogen atom had been introduced into the fourteen-membered ring.[30] The compound, which was subsequently developed

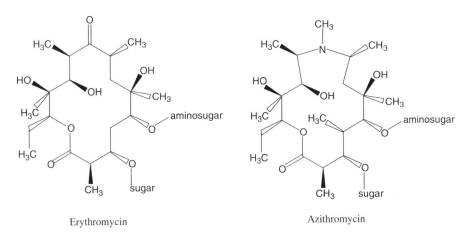

Erythromycin   Azithromycin

Figure 9   Erythronolides.

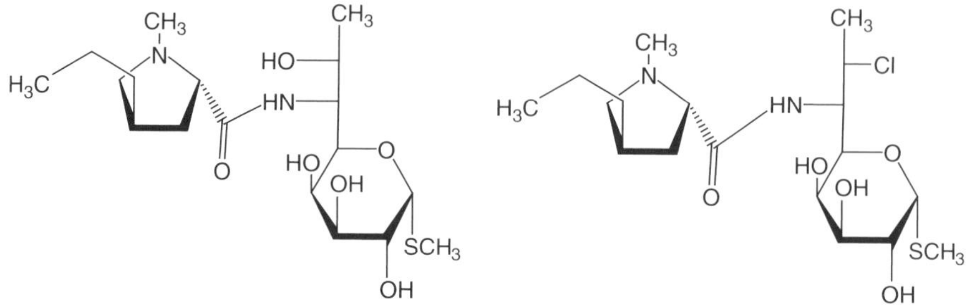

Lincomycin   Clindamycin

Figure 10 Lincomycin and clindamycin.

under license by Pfizer, proved significantly more stable that the parent. It provided long-lasting blood levels that made once per day treatment feasible.

A soil sample from a much less exotic source led to another new class of drugs. A detail man in the American Midwest sent a sample from Lincoln, Nebraska, back to his employer, the Upjohn Company, in Kalamazoo. There, scientists isolated an antibiotic whose chemical structure and range of activity was markedly different from the other hitherto known agents. It was named lincomycin (Fig. 10) after the producing organism, *Streptomyces Lincolnensis*, which in turn had been named after the capital of Nebraska.[31] The drug was eventually largely superseded by the derivative clindamycin, which was produced by modifying the structure chemically.

The forgoing touches on only the highlights of what was a major program in many laboratories from the late 1940s on. The screening process became highly efficient with time. In brief, this involved culturing soil samples so as to encourage formation of colonies of native fungi. Samples of the fungi would then be streaked onto plates inoculated with some test bacterium. The finding of clear kill zones around the colony would start an involved process for characterizing the substance that had killed the test bacteria. An unexpected major stumbling block in this search was the frequent rediscovery of previously known antibiotics. Many laboratories thus maintained extensive dictionaries of the properties of all antibacterial substances produced by soil organism. These tomes, maintained in order to avoid wasting time on known substances, were considered highly confidential and were jealously guarded, particularly on occasions where an employee left the company.

Penicillin had in the meantime not been forgotten. Limited work on attaching unusual acids to the beta-lactam part of the molecule by feeding those to the fermentation tanks had not been particularly successful. In 1959, however, scientists at Beecham Laboratories in the UK managed to devise conditions that allowed them to isolate the beta-lactam itself without the attached acid part from fermentation broths.[32] The availability of this substance, called 6-aminopenicillanic acid, or more familiarly 6-APA, offered organic chemists the chance to hook acids (also called side chains) that are never found in nature onto the active part of the molecule. The first drug from this research, pheneticillin, was developed by Beecham in collaboration with Bristol Myers in Syracuse, New York. This development was eagerly seized upon by a number of competing laboratories. Many of these now launched their own programs aimed at synthesizing and testing analogs of penicillin with novel side chains (Fig. 11). The output of these penicillin analogs was limited only by the high skill needed to carry out manipulations on these sensitive and reactive molecules. The work was subsequently facilitated by new methods for converting the much more easily available penicillin-V into 6-APA by either chemical or enzymatic means.[33] These new so-called semi-synthetic penicillins included much more stable drugs as well as a number with broader activity and some that were active when taken by mouth. This massive research effort resulted in at least 35 discrete substances that showed enough promise that their sponsors went through the process of acquiring nonproprietary names. Not all, needless to say, made it to the market. The orally active semi-synthetic drug amoxycillin is one of the more widely prescribed antibiotics to date. The beta-lactam pirbencillin is typical of some of the more highly modified penicillin-based drugs.[34]

The development of resistance to antibiotics on the part of bacteria can be considered a simple expression of evolution, fitness in this case being expressed as sheer survival. This phenomenon was probably first observed with penicillins because of their early widespread and indiscriminate use. The strained four-membered beta-lactam ring in these compounds is at one and the same time the

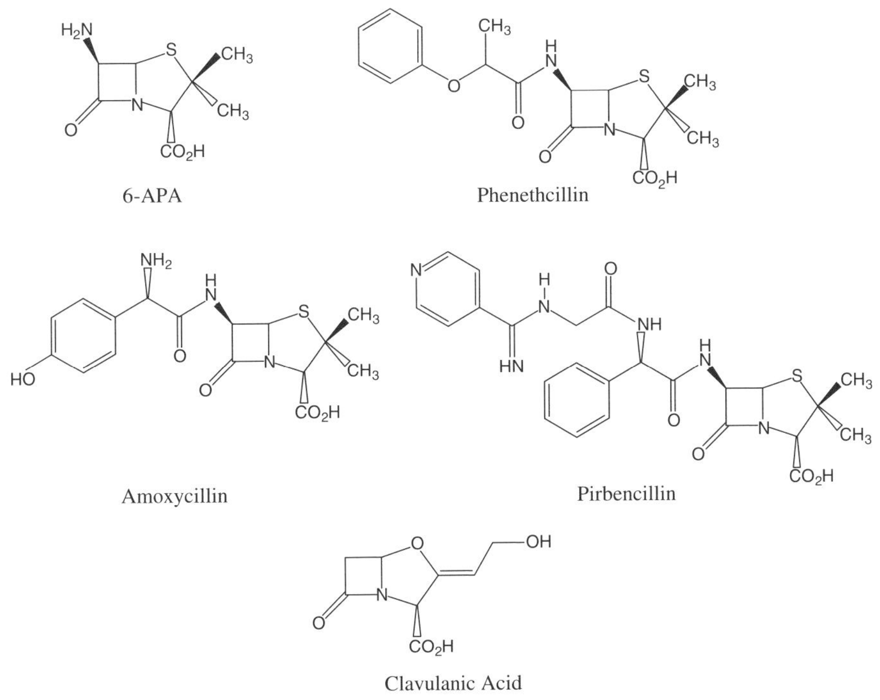

6-APA

Phenethcillin

Amoxycillin

Pirbencillin

Clavulanic Acid

Figure 11   Penicillins.

mode of action and Achilles heel. Resistant organisms elaborate an enzyme, beta-lactamase, that specifically inactivates that function. A fermentation product closely related structurally to the penicillins was found to compete for beta-lactamase with penicillins. This compound, clavulanic acid, which has little if any antibacterial activity in its own right, will thus retard inactivation of co-administered antibiotic.[35] Augmentin®, a fixed combination with amoxycillin, is widely prescribed for many infections.

Other environments rich in bacteria and fungi were examined for potential antibiotics as well. The lead for a new series of drugs based on a beta-lactam came from the isolation, in 1945, of a mixture of antibiotics from Sardinian sewage sludge by the Italian scientist Brotzu.[36] The active principle, named Cephalosporin C after the producing species, *Cephalosporium acremonium*, was too weakly active to be considered as a drug candidate. However, some of its properties, such as resistance to beta-lactamase, a bacterial enzyme that destroyed the four-membered ring in penicillin, made it an attractive starting point for further research. Although the chemical structure differed from penicillin, it shared enough similar features, such as the beta-lactam part, to lead scientists to apply the same methodology to try to prepare more active compounds. Attempts to introduce different side-chain acids by adding those to the fermentation were not very successful, nor were experiments aimed at producing the bare beta-lactam part of the molecule, equivalent to 6-APA, by fermentation. A procedure for obtaining the bare nucleus, 7-ACA (7-aminocephalosporanic acid), from cephalosporin C either chemically or by treatment with enzymes was finally published in 1962 by a team of chemists at Ciba in Basel (Fig. 12).[37]

The availability of this intermediate now made it possible to launch research programs analogous to those that had led to the collection of semi-synthetic penicillins. The group at Eli Lilly was particularly active in this field. Its first drug from this program was the injectable antibiotic cephalothin (Keflin). The starting material for this and later products, Cephalosporin C, was more difficult to obtain than the penicillins. Extensive research on production methods combined with demand had drastically lowered the price of bulk penicillins. This had in fact become a bulk commodity chemical, which, by the late 1990s, could be bought for less than one dollar per gram. The fact that both molecules shared a beta-lactam led to

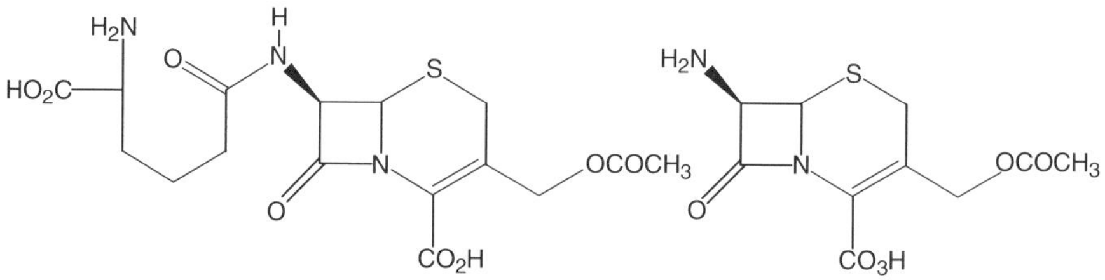

Cephalosporin C   7-ACA

Figure 12 Cephalosporin C and 7-ACA.

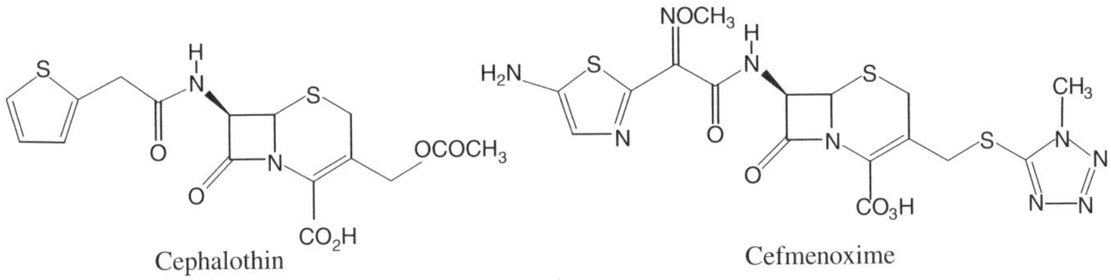

Figure 13   Cephalothin and Cefmenoxime.

considerable research on the part of chemists to find a way to transform penicillins into cephalosporins. These efforts were rewarded when Robert Morin devised just such a procedure at the Lilly labs. It is likely that the great preponderance of today's cephalosporin drugs start life as penicillin V or possibly its cousin penicillin G. The now ready availability of 7-ACA led to intensified research throughout pharmaceutical company laboratories. This resulted in a virtual flood of antibiotics, of which no fewer than 44 of these were assigned nonproprietary names. The group of drugs that were made available to physicians included many that could be administered orally. Most were quite resistant to the bacterial enzyme that destroys the beta-lactam. One of the most significant advances lay in the fact that selected semi-synthetic cephalosporins were active against a great many types of bacteria that were not sensitive to the earlier drugs. The beta-lactam nucleus is barely perceptible in the structure of some of the so-called third-generation antibiotics derived from 7-ACA such as cefmenoxime (Fig. 13).

Activity of the cephalosporins suggested that the nature of the ring fused to the beta-lactam was not crucial. As chemical methods and expertise accrued, it became possible to make modifications to that fused ring. A number of drugs that included such modifications have been approved for clinical use. Probably the most complex is moxalactam, a cephalosporin-like compound in which oxygen replaces sulfur (Fig. 14).[38] The involved lengthy synthesis of this compound at one stage involves degrading penicillin V to a bare four-membered ring. Chemists at Shionigi then built the oxygen-containing ring onto that. Imipenem represents another drastically modified compounds.[39] One approach to creating this compound involves an intermediate obtained by degrading the fused five-membered ring in 6-APA; it is

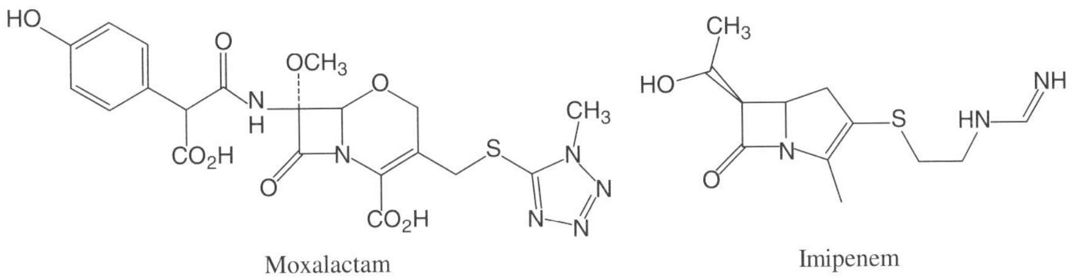

Figure 14   Beta-lactams fused to carbo- and oxacylic rings.

then built it up again with a ring in which sulfur has been replaced by a carbon atom.[40] The resulting antibiotic is active against a particularly wide assortment of bacteria. It is usually administered with cilastatin, a compound that slows metabolic destruction of the antibiotic.

The beta-lactam ring in all antibiotics of this class discussed thus far has been invariably fused onto another ring. The diversity of structures generated by that appendage would seem to indicate that this is not a critical element for biological activity. Nature soon demonstrated that this extra ring could be omitted entirely. The basic premise for screening soils as a source for antibiotics relies on the assumption that fungi and actinomycetes carve out a niche in the environment by secreting substances that will eliminate encroaching bacteria. This approach was soon plagued, as noted earlier, by rediscovery of known substances and lack of new findings. Attention then turned to bacteria themselves as a potential source for new antibiotics, because it was logical to assume that different microbial species would compete for the same ecological niches. In 1982, scientist at Squibb announced the discovery of a novel antibiotic produced by the bacterium *Chromobacterium violacium*; the active compound was dubbed a monobactam as it lacked a second ring fused onto the beta-lactam (Fig. 15).[41] The structure of this compound differs from the previously known antibiotics in this class not only by the lack of the fused ring but also by the lack of a complex organic acid side chain. An unusual sulfonic acid amide (NSO_3H) takes the place of the carboxylic acid (CO_2H) that is attached to the fused ring of the two-ring compounds. As is more often the case than not, the naturally produced antibiotic was not suitable for use as obtained. Its strong resistance to beta-lacatamase prompted chemists to undertake programs aimed at producing synthetic monobactams. The first compound that resulted from this program, aztreonam, proved, as predicted from its natural forerunner, to be particularly resistant to beta-lactamases. The synthesis of aztreonam starts with

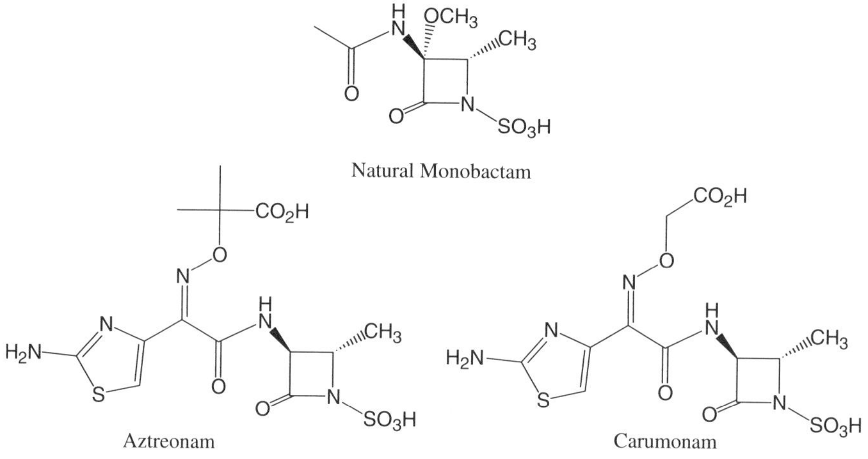

Natural Monobactam

Aztreonam

Carumonam

Figure 15   Single-ring beta-lactams.

simple organic chemicals.[42] This is, of course, in contrast to most modified beta-lactams, whose production involves modification of fermentation-derived starting materials. A parallel program at Takeda led to the commercial monobactam carumonam.[43]

Pharmaceutical research in the early 1960s relied heavily on the output of its organic chemists. These individuals as a rule synthesized compounds that were aimed at some particular therapeutic target and were tested in a model for that disease. Surplus amounts of the samples were then tested in a standard screen. This usually comprised an array of tests that were designed to identify chemicals that showed some degree of biological activity in other disease models. The rationale for this procedure lay in the observation of the number of compounds that had been found over the years to have more activity on some unintended endpoint than their purported target. Some of the compounds discovered by the screening procedure had provided leads for agents that had gone on to become successful drugs. In that period, the Sterling Drug laboratories had a long-standing program on tropical disease. A compound, submitted by one of their chemists, George Lesher, showed unexpected activity in the ongoing antibacterial screen. According to one account this was a byproduct from the synthesis of an antimalarial drug.[44,45] The chemical showed enough activity for further development. This drug, nalidixic acid (Fig. 16), was developed further and approved for use in treating urinary tract infections. The chemical structure of this little heralded antibiotic, it should be noted, incorporated all the necessary features that would be found in the later quinolones. The activity of this forerunner was quite modest and was apparently not enough to stimulate research in competing laboratories. Only a few additional

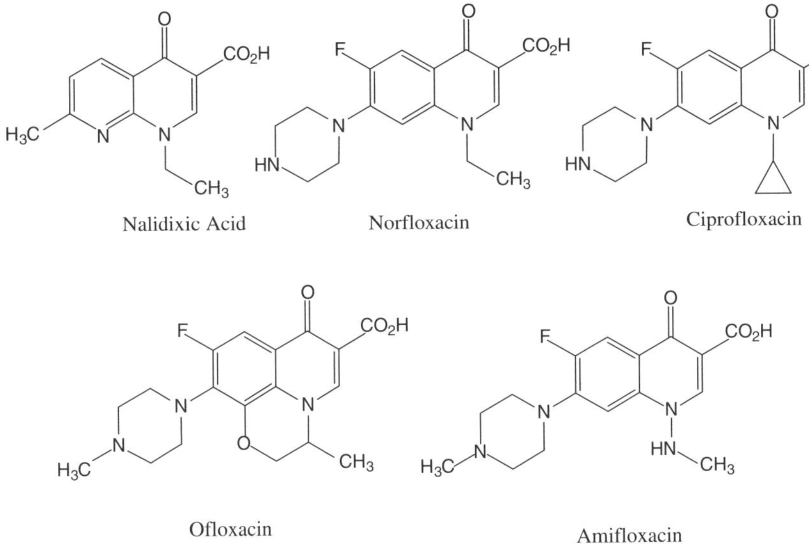

Figure 16   Naldixic acid and current quinolone antibiotics.

entities reached the clinic from an assortment of other firms over the next decade. These analogs were only modestly more effective than the forerunner. Research did apparently continue at a low level in several laboratories. Replacement of hydrogen by fluorine was known to markedly increase activity in several classes of drugs such as the corticosteroids. Two specific modifications of the chemical structure of the quinolones, one of which interestingly comprised introduction of a fluorine atom, dramatically changed the picture. This compound, norfloxacin, was first synthesized in the late 1970s in the laboratories of the Kyorin Pharmaceutical Company in Japan.[46] Its activity in test models and subsequently in humans was clearly superior to any of its predecessors. It was then introduced into the United States by Merck. The excellent broad spectrum activity of this novel drug stimulated a virtual race to produce new chemical entities and it has been estimated that over a thousand quinolones have been synthesized in various laboratories, 22 of which have been assigned nonproprietary names. Although the structures of some of these later entities, such as ofloxacin and amifloxacin, have become quite elaborate, they still incorporate the basic elements present in the lead compound. Another of the second-generation congeners, ciprofloxacin, known to the press as simply cipro, gained considerable fame in late 2001. This drug gained wide exposure as a remedy and preventative at the height of the anthrax scare.

The quinolones, it has been determined, act by a unique mechanism. Recall that DNA is a very large string-like molecule. The length of this compound must be reckoned in feet rather than millimeters. This entity must obviously be tightly coiled for one to be packed into each and every cell nucleus. It would take far too long to carefully uncoil what has been compared to a snarl of fishing monofilament in order to read a stretch of DNA to exert its controlling role. Biochemistry then does what any impatient fisherman would do: cut sections in order to pass strands through another. A very special enzyme called topoisomerase then temporarily holds the cut ends together until the section has passed through and then subsequently reconnects the cut pieces. Bacterial DNA floats loose within the cell as microbes have no cell nucleus. Their topoisomerase, possibly because of this difference, is quite different from that in organisms that have a cell nucleus. The quinolones act as quite specific inhibitors of bacterial topoisomerase and in effect degrade the DNA as the cut ends are not identified for reconnection. The lack of an intact DNA template prevents the affected organism from replicating.[47,48]

The very large selection of very effective antibiotics has led to a significant diminution of research on new antibiotics. The increasing incidence of antibiotic-resistant organisms has led to renewed attention in the field. A drug introduced within the past five years may be an early harbinger of this. The development of this agent begins with the discovery of a series of compounds at the Du Pont Laboratories, which showed good activity against a wide range of bacteria.[49] The most promising candidate, DuP 721, was particularly active against microbes resistant to other classes of drugs. It was not taken further because of safety concerns raised from results in preliminary animal toxicity tests. Following these reports, scientists at Upjohn launched a program to prepare related compounds. All the

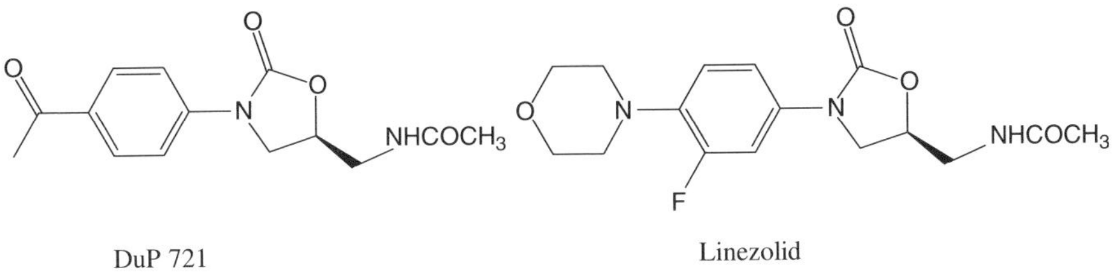

Figure 17   Linezolid.

analogs contained the nitrogen and oxygen five-membered ring, called an oxazoli-done, also present in the Du Pont series. Two of the compounds from the collection of new analogs synthesized at Upjohn showed very promising antibacterial activity.[50] One of these, linezolid, which has been approved for human use, is quite effective against classes of bacteria resistant to known antibiotics (Fig. 17). This drug also acts by inhibiting bacterial protein synthesis. It does so by a rather unique mechanism, by blocking the binding of the transfer RNA intermediate to the unit that adds amino acids to the protein chain. This in essence stops the process before it even starts. It was felt that this differs sufficiently from the way other antibiotics act, by inhibiting bacterial protein, to avoid the development of cross-resistance.[51]

The various cell-killing substances uncovered by the soil-screening programs led a number of laboratories to examine some of those extracts for their potential for uncovering antineoplastic agents.[52] These programs were based more on the availability of large numbers of extracts than on any imagined similarity between antibiotic and anticancer activity. The former of course relies on the fact that bacteria and humans belong to different kingdoms. The highly selective toxicity to bacteria is due to interference with biological mechanisms simply not present in humans. No such marked differences unfortunately obtain between normal and cancer cells. Most current antineoplastic agents thus rely largely on inhibiting processes that occur at a higher rate in cancer cells. The majority of these drugs act to inhibit cell replication at the level of DNA by a variety of mechanisms. This might be called the kinetic approach to selective toxicity. The antitumor antibiotics have not, with the exception of the anthracylines, led to classes of agents with related structures. Although considerable work may have been devoted to developing more effective and better tolerated representatives, they are, as a rule, represented by a single licensed representative.

The simplest of these agents is the modified sugar streptozotocin (Fig. 18) isolated from a *Streptomyces* fermentation broth. The structure consists of glucose in which one of the hydroxyl groups has been replaced by a very reactive nitrosourea group.[53] It is of interest that the very same grouping occurs in the synthetic anti-cancer drugs BCNU and CCNU. These are some of the very first anticancer drugs; they act as alkylating agents on DNA and thus prevent cell replication. Streptozotocin is somewhat selectively taken up by the pancreas. Its main indication is thus for treatment of pancreatic cancer. This selective uptake and consequent

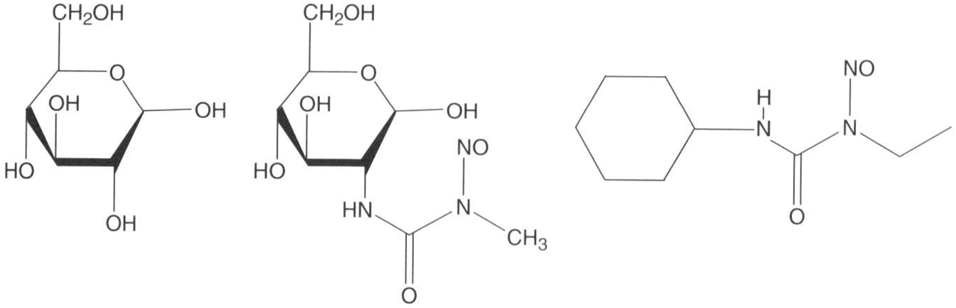

| Glucose | Streptozotocin | CCNU |

Figure 18 Streptozotocin and CCNU.

toxicity has provided pharmacologists with an important tool. Administration of the drug to laboratory animals causes them to develop diabetes. So-called streptozocin rodents are consequently often used in diabetes research.[54]

Mitomycin C (Fig. 19) is a somewhat more complex alkylating agent isolated from another *Streptomyces* strain.[55] As is often the case, the compound is accompanied by the closely related mitomycins A and B. This agent is among the drugs that must first undergo a transformation in order to exert its action. The drug seems to be most often used in combinations with other antineoplastic agents in the various combinations designated by acronyms. It is the M in MIP (mitomycin C, ifosfamide, and cylophosphamide). Although considerable work has been devoted to producing related compounds in the search for more selective and less toxic analogs, it remains the only representative of its class approved for use in the United States.

Three structurally very complex fermentation products, plicamycin,[56] dactinomycin, and bleomycin have also found their place in oncology (Fig. 20). Each, it should be added, represents a single drug class. The structure of the first of

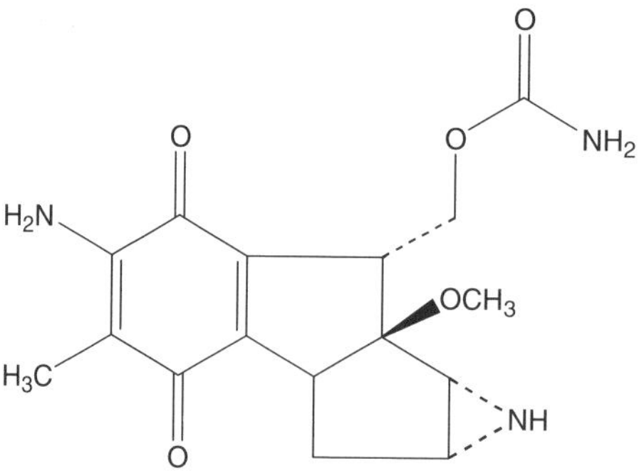

Mitomycin C

Figure 19 Mitomycin C.

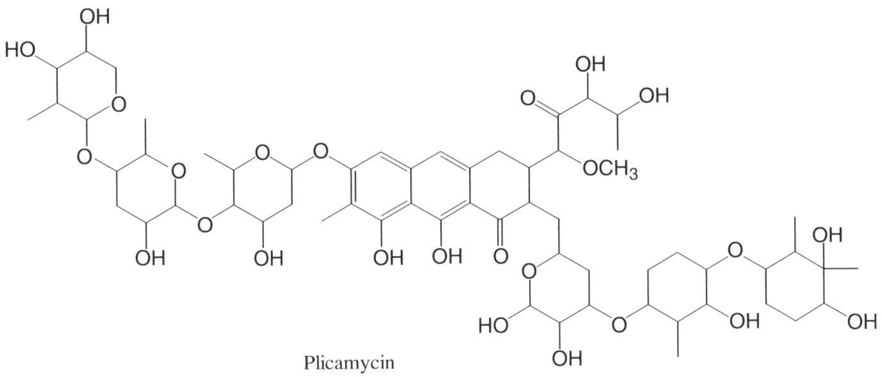

Plicamycin

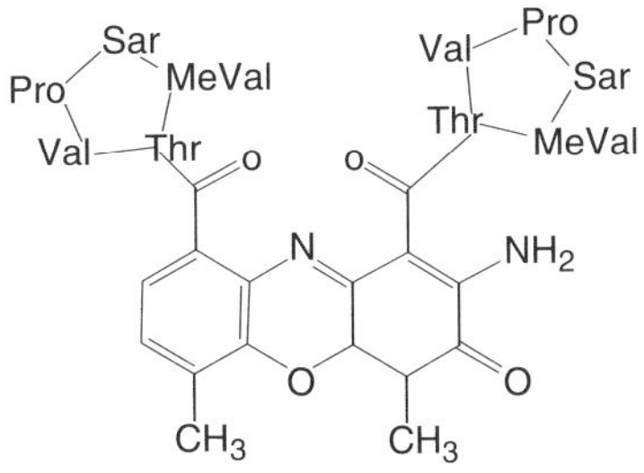

Dactinomycin

Figure 20   Plicamycin and dactinomycin.

these compounds has several structural features, such as the linear array of rings bearing hydroxyl groups and the attached sugars, which foreshadows the currently widely used anthracylines. Plicamycin, initially known as aureolic acid, is apparently used mainly for treating testicular cancer. Dactinomycin, also known as actinomycin D, is one of yet another group of closely structurally related compounds isolated from *Steptomyces* fermentation broths.[57] This compound too has been the subject of a large amount of research aimed at producing a better congener. It is of interest that its mechanism of action is similar to the anthracyclines that will be considered next. The structure of bleomycin is not shown.

FDA approval of adriamycin in the early 1980s marked a significant change in cancer chemotherapy. The wide spectrum of activity of this drug against a variety of tumors led to increased recourse to chemotherapy for treating this disease. The development actually started at the Farmitalia laboratories in the 1950s with the isolation of a red pigment from cultures of *Streptomyces peucitius*. (The name is said to derive from the Peucitia region in Italy.) Over the years, Arcamone and his colleagues at the company isolated a number of pure substances from the mixture that made up the pigment. The first of these was a compound with antibiotic activity that they named daunomycin (Fig. 21). The same compound was isolated in parallel work at Rhône-Poulenc in France, where it was dubbed rubidomycin.

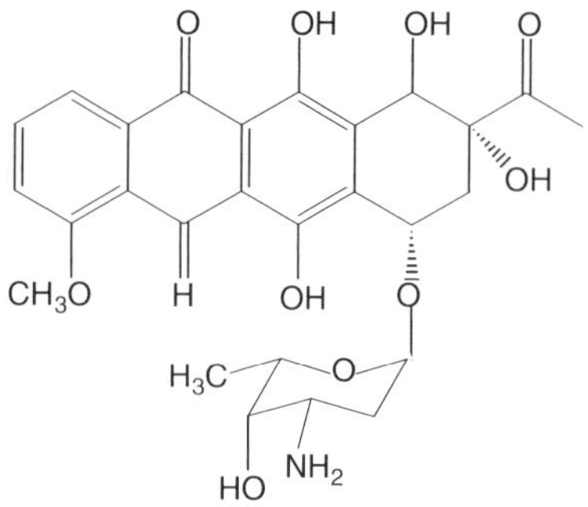

Dauonorubicin
Daunomycin

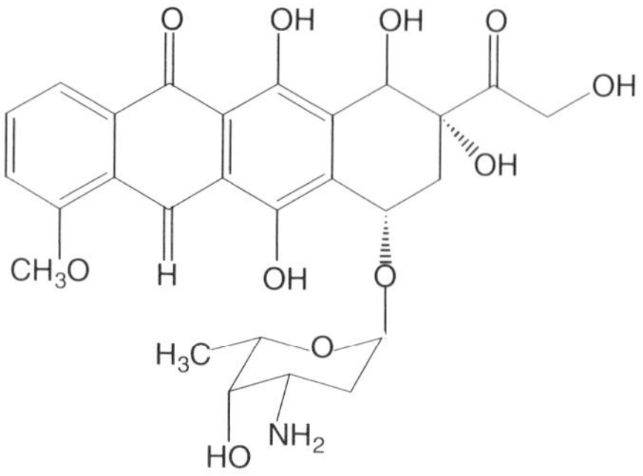

Doxorubicin
Adriamycin

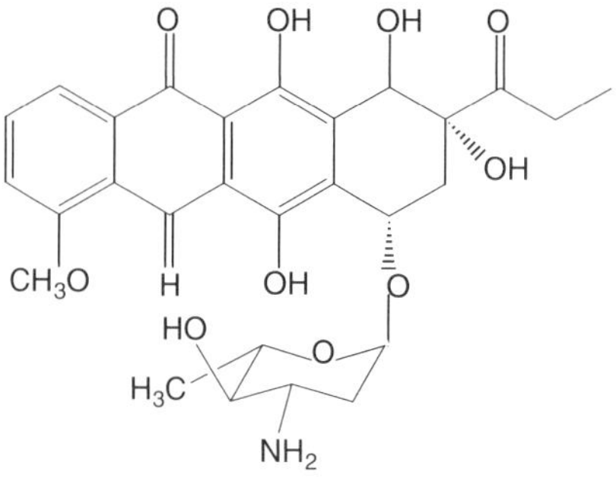

Epirubicin

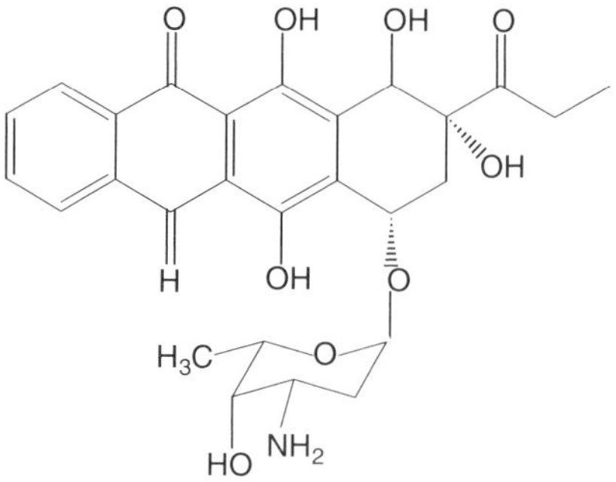

Idarubicin

Figure 21 Anthracycline antitumor antibiotics.

This compound was found to have antibiotic activity and, more importantly, to be active against leukemias. The nonproprietary name daunorubicin, which included syllables from each source, was assigned to the drug. The "rubicin" suffix was to be used for all subsequent members of the series. Daunomycin was eventually introduced clinically, where it found its place as part of combinations used to treat leukemias. Further work in Milan then led to the identification of the analog doxorubicin, which has an additional hydroxyl group on the side chain.[58] This compound is more familiarly known as adriamycin, the name deriving from the Adriatic Sea; it seemed to have much wider activity and found widespread use in oncology. As a side note the Farmitalia's American affiliate, Adria Laboratories, was actually named after the drug. Clinical use revealed a serious shortcoming of this drug. In addition to the usual toxicities, adriamycin was associated with dose-limiting cardiac toxicity. Chemists at Farmitalia started significant programs to modify various parts of the molecule to try to overcome that effect. Epirubicin, in which one of the hydroxyl groups on the sugar is inverted, is said to be less cardiotoxic, permitting a higher cumulative dose. Idarubicin, the analog missing a methyl ether on the first ring is not in fact a fermentation product; the compound was first produced by total synthesis.[59] The anthracyclines were at first thought to exert their cytotoxic effect by intercalation in DNA. The large flat molecule, it was thought, would slide between two hydrogen-bonded base pairs and thus disrupt DNA's organization. The disrupted molecule would then no longer be able to take part in cell replication. More recent results suggest that the compound also disrupts topoisomerase 2, one of the human enzymes involved in cell replication whose action is analogous to the bacterial topoisomerase disrupted by the quinolone antibiotics.

REFERENCES

1. Federal Interagency Forum on Aging and Related Statistics; available at http://agingstats.gov/tables%202001/tables-healthstatus.html.
2. S. GARFIELD, *Mauve*, W.W. Norton & Co., NY, 2001, pp. 155–157.
3. G. DOMAGK, *Deut. Med. Wochenschr.*, 61, 829 (1935).
4. J. & MME. TREFOUEL, F. NITTI AND D. BOVET, *C.R. Seances Soc. Biol.*, 120, 756 (1935).
5. E. FOURNEAU, J. & MME. TREFOUEL, F. NITTI AND D. BOVET, *C.R. Seances Soc. Biol.*, 122, 652 (1936).
6. L. F. FIESER, *Experiments in Organic Chemistry*, D.C. Heath & Co., Boston, 1955, p. 147.
7. A. FLEMING, *Br. J. Exp. Path.*, 10, 226 (1929).
8. E. LAX, *The Mold in Dr. Florey's Coat*, Henry Holt & Co., NY, 2004.
9. R. D. COGHILL AND R. S. KOCH, *Chem. Eng. News*, 23, 2310 (1945).
10. http://www.ars.usda.gov/is/timeline/penicillin.htm.
11. E. BRANDL AND H. MARGREITER, *Oesterr. Chem. Ztg.*, 55 (1954).
12. http://www.ahc-net.at/0001/antibiotika_monitor/56_01/56_01_6.htm (in German).
13. J. T. PARK AND J. L. STROMINGER, *Science*, 125, 99 (1957).
14. D. LEDNICER AND L. A. MITSCHER, *The Organic Chemistry of Drug Synthesis*, Wiley, NY, 1977, pp. 120–130.
15. R. TSCHESCHE, *Arzneimittelforschung.*, 1, 335–339 (1951).
16. S. R. BUSHBY AND G. H. HITCHINGS, *Br. J. Pharmacol.*, 33, 72 (1968).
17. G. H. HITCHINGS, *J. Infect. Dis.*, 128, Suppl. 433 (1973).
18. H. MASKE, *Med. Klin. (Munich)*, 50, 2083 (1955).
19. D. LEDNICER AND L. A. MITSCHER, *The Organic Chemistry of Drug Synthesis*, Wiley, NY, 1977, p. 136.
20. For a discussion of sulfonamide diuretics see R. C. ALLEN, in E. J. CRAGOE Jr., Ed., *Diuretics*, Wiley, NY, 1933, pp. 49–200.
21. S. A. WAKSMAN, *Br. Med. J.*, 2, 595 (1950).
22. M. WAINWRIGHT, *Hist. Philos. Life Sci.*, 3, 97 (1991), for the controversy on the inventorship of streptomycin.
23. H. KAWAGUCHI ET AL., *Antibiot.*, 25, 695 (1972).
24. M. J. WEINSREIN ET AL., *Antimicrob. Ag. Chemother.*, 1 (1963).
25. L. GOZALES AND J. P. SPENCER, *Am. Fam. Physician*, 58, 1811 (1998), for a review of clinical applications.
26. B. M. DUGGAR, *Ann. N. Y. Acad. Sci.*, 51, 177 (1948).
27. C. R. STEPHENS, R. K. BLACKWOOD AND M. S. VON WITTENAU, *J. Am. Chem. Soc.*, 85, 2643 (1963).
28. P. E. SUM AND P. PETERSEN, *Bioorg. Med. Chem. Lett.*, 1459 (1999).
29. K. R. GERZON, E. H. FLYNN, M. V. SIGAL, P. F. WILEY, R. MONAHAN AND U. C. QUARCK, *J. Am. Chem. Soc.*, 78, 9396 (1956).
30. S. DHOKIC, N. LOPATAR, G. KOBREHEL, H. KRNJEVIC AND O. CAREVIC, U.S. Patent 4,886,792 (1989).
31. D. J. MASON AND C. LEWIS, *Antimicrob. Ag. Chemother.*, 10, 7 (1964).
32. F. R. BATCHELOR, D. GAZZAED, J. H. C. NAYLER AND G. N. ROLINSON, *Nature*, 183, 257 (1959).
33. G. N. ROLINSON, *J. Antimicrob. Chemother.*, 41, 589 (1998).

34. For detailed accounts of the related drugs see the series D. LEDNICER AND L. A. MITSCHER, *The Organic Chemistry of Drug Synthesis* (V1, 1977; V2, 1978; V3, 1984; V4, 1990) and D. LEDNICER, *The Organic Chemistry of Drug Synthesis* (V5, 1995; V6, 1999). Wiley, NY, NY.

35. D. G. BRENNER AND J. R. KNOWLES, *Biochemistry*, 23, 5833 (1984).

36. M. NARISADA ET AL., *J. Med. Chem.*, 22, 757 (1952).

37. I. SHINKAI, R. A. REAMER, T. LIU, K. RYAN AND M. SLETZINGER, *Tetrahedron Lett.*, 21, 2783 (1982).

38. For a review of the discovery of cephalosporin, see J. M HAMILTON-MILLER, *Int. J. Antimicrob. Ag.*, 20, 179 (2000).

39. M. GORMAN, R. R. CHAUVETTE AND S. KUKOLJA, in J. S. BINDRA and D. LEDNICER, Eds, *Chronicles of Drug Discovery*, Vol. 2, Wiley, NY, 1983, p. 49 et. seq.

40. For a summary of highly modified beta lactams see D. LEDNICER, *Strategies for Organic Drug Synthesis and Design*, Wiley, NY, 1998, pp. 407 et. seq.

41. J. S. WELLS, W. H. TREJO, P. A. PRINCIPE, K. BUSH, N. GEORGOPAPADAKON, D. P. BONNER AND R. B. SYKES, *J. Antibiot.* (Tokyo), 35, 184 (1982).

42. C. M. CIMARUSTI, in D. LEDNICER, Ed., *Chronicles of Drug Discovery*, ACS Books, Washington, DC, 1993, p. 239.

43. S. HASHIGUSHI, Y. MAEDA AND S. OCHIAI, *Heterocycles*, 24, 2273 (1986).

44. www.baytril.com/6/History.htm.

45. E. W. MCCHESNEY, E. J. FREOLICH, G. Y. LESHER, A. V. CRAIN AND D. ROSI, *Toxicol. Appl. Pharmacol.*, 61, 292 (1964).

46. http://www.kyorin-pharm.co.jp/eg/act/soyaku.html.

47. J. T. SMITH, *J. Antimicrob. Chemother.*, 18, Suppl D, 21 (1986).

48. A. L. BARRY, *Antimicrob. Ag. Chemother.*, 32, 150 (1988).

49. W. A. GREGORY, D. R. BRITTELI, C. I. WANG, M. A. WUONOLA, R. J. MCRIPLEY, D. C. EUSTICE, V. S. EBERLY, P. T. BARTOLOMEW, A. M. SLEE, AND M. FORBES, *J. Med. Chem.*, 32, 1673 (1989).

50. S. J. BRICKNER, D. K. HUTCHINSON, M. R. BARBACHYN, P. R. MANNINEN, D. A. ULANOWICZ, S. A. GARMON, K. C. GREGA, S. K. HENDGES, D. S. TOOPS, C. W. FORD AND G. E. ZURENKO, *J. Med. Chem.*, 39, 673 (1996).

51. D. SHINABARGER, *Expert Opin. Investig. Drugs*, 1195 (1999).

52. W. T. BRADNER and C. A. CLARIDGE, in W. A. REMERS, Ed., *Antineoplastic Agents*, Wiley, NY, 1984, p. 41 et. seq. for a discussion of the screens then in use.

53. M. E. HERR, H. K. JAHNKE AND A. D. ARGOUDELIS, *J. Am. Chem. Soc.*, 89, 4808 (1967).

54. S. S. KATYARE AND J. G. SATAV, *Diabetes, Obesity and Metabolism*, 7, 555 (2005).

55. S. WAKAKI, *Cancer Chemother. Rep.*, 13, 79 (1961).

56. Y. U. BERLIN, O. A. KISELVA, M. N. KOLOSOV, M. M. SHEMYAKIN AND V. S. SOIFER, *Nature*, 218, 193 (1968).

57. C. E. DALGLIESH AND A. R. TODD, *Nature*, 164, 820 (1949).

58. See F. ARCAMONE, Ed., *Doxorubicin Anticancer Antibiotics*, Academic Press, NY, 1981, for a review of the development and chemistry of the anthracyclines.

59. F. ARCAMONE, L. BERNARDI, B. PATELI, P. GIARDINO, A. DIMARCO, A. M. CASAZZA, C. SORANZO AND G. PRATESI, *Experientia*, 34, 1255 (1970).

Chapter 2

Antiviral Drugs

The discovery of antibiotics, as noted in the previous chapter, can be traced back to the use of dyes used to visualize bacteria. Left unsaid in that discussion was the fact that the growth of microbiology was intimately connected to the ability to visualize the organism in question by means of microscopes. The discovery of viruses in essence depended on negative evidence. As ever more pathogenic organisms were discovered, bacteriologists tended to attribute the cause of most diseases to bacteria. Microbiologists in the late nineteenth century who were studying tobacco mosaic disease, however, sought in vain for the microbes that were at the time considered the probable causative organism of that infection. In 1892 the Russian scientist Dimitry Ivanovsky prepared a solution taken from tobacco leaves from a sick plant and passed that through an extremely fine ceramic filter. He then showed that exposing plants to the filtrate would cause the disease in new plants. The fact that the pores in the filter should have excluded bacteria pointed to a smaller entity as the cause for tobacco mosaic disease. The pores of the filter were fine enough to preclude any chance that unknown agent could be visualized. Within a few years it was recognized that other plant, animal, and even human diseases for which no bacterial cause could be identified could be traced to similar agents. As early as 1900, Walter Reed, for example, was able to point to a nonbacterial cause of yellow fever by transmitting the disease among mice.

Early work on these agents, soon named viruses, was plagued by the inability to culture them; viruses would simply not replicate in any of the traditional media that had proven so successful in bacteriology. Viruses, it was found, would only propagate in a living organism. Tobacco mosaic, for example, required living leaves. By this means Wendel Stanley was able in 1935 to isolate and crystallize tobacco mosaic. The inability to propagate viruses in the absence of living cells pointed to the nature of the agent and at the same time presaged the problems associated with finding drugs for treating viral disease. By the 1930s, various methods had been developed for overcoming the problem. These techniques all involved methods for propagating viruses that used living cells such as those in fertilized eggs and the newly developed cell cultures.

Jumping ahead to the present, viruses are now recognized to comprise a series of agents that consist of relatively short segments of DNA and RNA encapsulated in a coat made up of proteins that may have attached sugars and fats. A compelling argument can be made for the proposition that viruses are not in fact living organisms. These very complex structures, called virions before they enter a cell, lack the ability to perform most functions associated with life, notably metabolism and reproduction. After penetrating a cell on a live host, a virus proceeds to hijack the cell and then causes it to first produce the special enzymes that the virus requires for its own life processes. The host cell is then induced to produce more virus particles. This process eventually kills the host cell as it can no longer fulfill the functions required for its own survival; some viruses eventually lyse the co-opted cell in order to allow fresh virus to escape.

The genetic material at the core of a virion can consists of either single- or double-stranded RNA or DNA. The highly organized protein capsule gives virions characteristic shapes when viewed with an electron microscope. A number of virions, such as that which causes HIV, have specific structures on the protein coat that bind to corresponding structures on the immune system cell that will be infected. Other virions simply bind with the surface of their targets. After the virion is engulfed by the cell, it needs to shed its protein coat. DNA viruses at this point simply cause the cells to translate that to complementary RNA to use that as a template to make fresh, now viral proteins. Some RNA-containing virions cause the cell to make viral proteins using that directly. The so-called retroviruses, on the other hand, call on an enzyme called reverse transcriptase to make complementary DNA, which then enters the process as though it were the cell's own genome. The captured cell then not only synthesizes new viral DNA or RNA cores, but also the proteins that will coat the new virions. Those proteins, as is often the case, form but a small segment of a much larger molecule. The needed segment is excised from that first-synthesized protein by special enzymes called proteases.

The efficacy and selectivity of antibiotics is due in great part to differences in the biochemical functioning between bacteria and humans. Those differences all relate to the means by which individual cells, be they host or bacterial, grow and replicate, in other words carry on a life cycle. Developing drugs that arrest viruses is commensurately more difficult because, as already noted, there is an ongoing discussion as to whether virions can be deemed to be alive. Selective killing agents thus depend on the very subtle differences in functions that exist on the place where viral replication occurs: that is, between normal cells and those infected by a virus. The drugs that have been developed attack one or another of the steps involved in the process of forming new virions.

The great majority of viral diseases, for example polio, influenza, and yellow fever, are in fact dealt with prophylactically with vaccines before infection sets in. This greatly diminishes the need to develop drugs that will attack the causative viruses. This approach started in the mid-eighteenth century with the use of cowpox, when Jenner introduced vaccination with cowpox as a means of preventing small-pox, the closely related human disease. This of course long pre-dates the discovery of viruses or, for that matter, bacteria. Vaccination involves administration of either attenuated or killed virions or even live ones from a closely related disease. The host's immune system will then hopefully manufacture antibodies that recognize and neutralize disease-causing virions when they are next met. Vaccines are not suitable for treating active infection because of the time it takes for antibodies to develop. The slow progress in developing antiviral drugs is due in part to the fact that work had to wait for the detailed knowledge and methods now to hand. The availability of vaccines to deal with some of the most threatening viral diseases may have played a part as well. Most of the early antiviral drugs date to the 1970s. It is of note, however, that over half of the drug classes now available were developed more recently in response to the HIV-AIDS epidemic. Many of those drugs are in fact intended to attack retroviruses; their utility is thus largely restricted to treating HIV.

By the 1960s, a number of pharmaceutical laboratories had in place in-vitro screens to test for antiviral activity compounds produced by their organic chemists. These high-throughput screens involved various techniques that relied on virus-infected cell cultures. One of those random screening programs led to the surprising finding that the adamantane derivative1-aminoadamantane (Fig. 1) displayed antiviral activity.[1] This activity was subsequently confirmed in the clinic. This compound, known by the nonproprietary name amantadine was introduced as Symmetrel®, presumably as a reflection of its chemical structure. The name adamantanes for this class of compounds derives from the fact that the first example was seemingly adamantly resistant to chemical manipulation. The parent hydrocarbon, adamantane, was originally isolated from petroleum in the Moravian province of Czecholsovakia in the early 1930s. Determination of chemical structures in those pre-instrument days depended largely on relation to compounds of known structures by chemical manipulation. The new compound resisted all reagents that the chemist Landa tried. The structure in the end was assigned on the basis of its molecular weight and ratio of hydrogen to carbon as determined by combustion analysis. Both this compound and rimantadine, a congener introduced later, have some utility in treating influenza A.[2] They are, however, without effect on other influenza

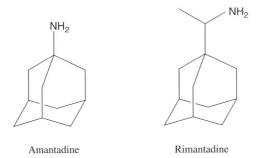

Amantadine   Rimantadine

Figure 1   Amantadine and rimantadine.

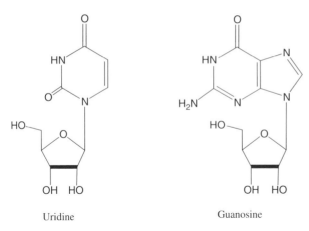

Uridine   Guanosine

Figure 2   Uridine and guanosine.

strains. Their common mechanism of action involves inhibition of the initial viral uncoating step.[3] The viral gene can, as a result, not be read by cellular enzymes. Amantadine has also found use in a quite unrelated disease. The drug seems to diminish tremors caused by Parkinson's disease.

Nucleosides such as the pyrimidine uridine (U) and the purine guanosine (A) comprise the essential building blocks of both DNA and RNA (Fig. 2). These compounds are in consequence intimately involved in cell growth and replication. Starting in the 1960s, many laboratories undertook synthesis programs to prepare modified versions of these compounds as potential antitumor agents. This was based on the rationale that the known faster turnover of cancerous as compared to normal cells would cause them to incorporate more of the modified nucleosides. Chemical features on the newly incorporated modified fragments would then prove fatal to the tumor cell. This program met with some success, providing several compounds still in use in cancer chemotherapy such as the pyrimidine 5-fluorouracyl and cytosine arabinoside (Ara-A).

Robins and his colleagues carried out an extensive synthesis program on modified nucleosides aimed at producing antitumor compounds. A number of these involved replacing the nitrogen-containing six-membered ring in nucleosides by other nitrogen-containing rings that had no counterpart in naturally occurring agents (Fig. 3). Many of the resulting products were tested for antiviral activity as well; one of those products, ribavarin,[4] had sufficient activity to be taken to the clinic. This compound is now approved for treating syncitial viral infections, a rare disease in infants, and for hepatitis C when combined with interferon. The analog edoxudine,[5] retains the six-membered ring, but it has an added ethyl group and a modified five-membered sugar. This drug has shown activity in genital herpes, although it has not been commercialized in the United States. The related analog sorivudine[6] also showed good activity against the herpes virus. It has been largely abandoned as a drug due to toxicity when combined with certain antifungal

Ribavarin Edoxudine Sorivudine

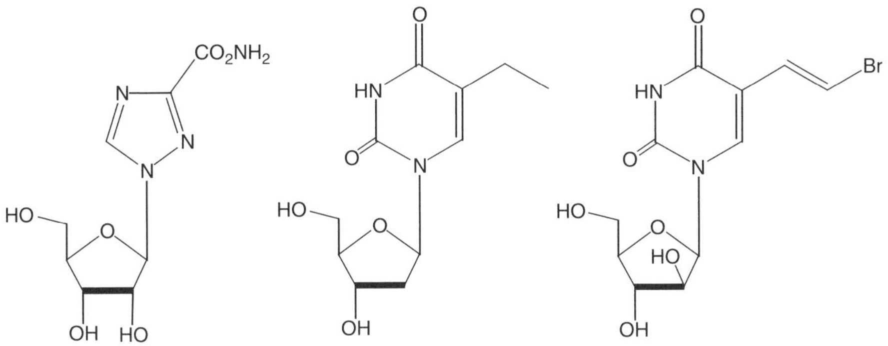

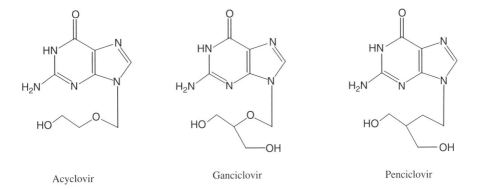

Figure 3 Ring-modified nucleosides.

agents. The exact mechanism by which these agents stop viral infections is still uncertain, but it is thought that incorporation in DNA or RNA may play a role.

A series of now widely used antiviral drugs have come from a modification in which the five-membered sugar is replaced by an open-chain equivalent. That portion of the molecule is depicted in the representatives in Figure 4 in admittedly tortured form to show the relation between the natural and "open" forms. The first of these compounds acyclovir, came from a long-standing program at the Burroughs Wellcome Labs on modified nucleosides and nucleotides. This compound in essence opens the ring by omitting one carbon atom and the oxygen atom thereon. The actual synthesis of course involves a quite different route.[7] The drug is active against a range of herpes viruses and is indicated for cold sores, shingles, and genital herpes. It is also useful for treating chicken pox as a result of its activity against the varicella virus. The later analog ganciclovir[8] restores to the sugar surrogate one of the hydroxyl groups present in the natural nucleoside. Ganciclovir is indicated for treating cytomegalovirus retinitis. The fused two-ring purine fragment in each of the two preceding agents is hooked to the side chain via a carbon atom

Acyclovir Ganciclovir Penciclovir

Figure 4 Open sugar ring nucleosides.

that is substituted by both oxygen and nitrogen. This is a somewhat reactive center that can facilitate degradation. The oxygen atom present in this compounds is replaced by a carbon atom in the analog penciclovir,[9] leading to what is probably a more stable compound. This agent carries much the same list of indications as acyclovir. These open-chain analogs, like the normal nucleosides, are not active per se, but must first be converted in the body to their triphosphate derivatives by stepwise addition of phosphate groups. The resulting triphosphates are then mistaken by the virus-infected cell for the guanine triphosphate required for building DNA and incorporated into the growing chain. The presence of the intruding nucleoside-like fragment causes premature chain termination and failure of new virion synthesis.

The appearance in the early 1980s, seemingly out of the blue, of a cluster of cases of a rare form of pneumonia caused by *Pneumocystis carinii* among homosexual men in San Francisco heralded the onset of the HIV-AIDS epidemic. The actual disease with which these men came down was found to be but a manifestation of the fact that the patient's immune system had been seriously depleted, hence the name Acquired Immune Deficiency Syndrome. It was established in relatively short order that the causative agent of the disease was a retrovirus that infected and in the end killed one of the first white cells that respond to a challenge to the immune system, the T4 lymphocytes. It was not yet recognized at that early stage that this would in a relatively short time blossom into a world-wide epidemic. Various political and social factors nevertheless set off a major effort in both government and private sector laboratories to find drugs for treating the disease as well as vaccines to prevent infection. The first goal has so far yielded at least four classes of compounds that arrest progression of the disease among individuals infected by the Human Immuno-deficiency Virus (HIV). The search for an effective vaccine has yet to deliver a widely effective agent.

The initial step in the replication of a retrovirus, recall, involves building a DNA strand in infected cells that reflects the viral genome. The specialized enzyme required for this task, reverse transcriptase, has proven a particularly fruitful target for drugs for treating HIV; no fewer that eight compounds in this class are currently approved (Fig. 5). The first of these, originally known by the trivial name azathymidine and better known as AZT, was in fact first synthesized in 1962 as part of the program to investigate nucleosides as potential antitumor agents.[10] This compound was one of the many candidates put through the newly developed anti-HIV screening programs. It was the first compound found to effectively protect cells in cultures against HIV with minimal toxicity to uninfected cells. There has been contention between the Burroughs–Wellcome and NCI-sponsored laboratories as to the priority of this discovery. Clinical trials on the compound, now assigned the nonproprietary name zidovudine, showed that the drug restored in good part the previously depressed levels of HIV patients' immune systems. With time, however, the virus developed resistance to the drug. This was partly overcome by combining AZT with drugs that work by other mechanisms. Zidovudine differs from thymidine, the nucleoside that is used to build DNA (T) only in that an azide ($N{=}N{=}N$) group replaces the hydroxyl of the endogenous compound. Incorporation of AZT into growing DNA by viral

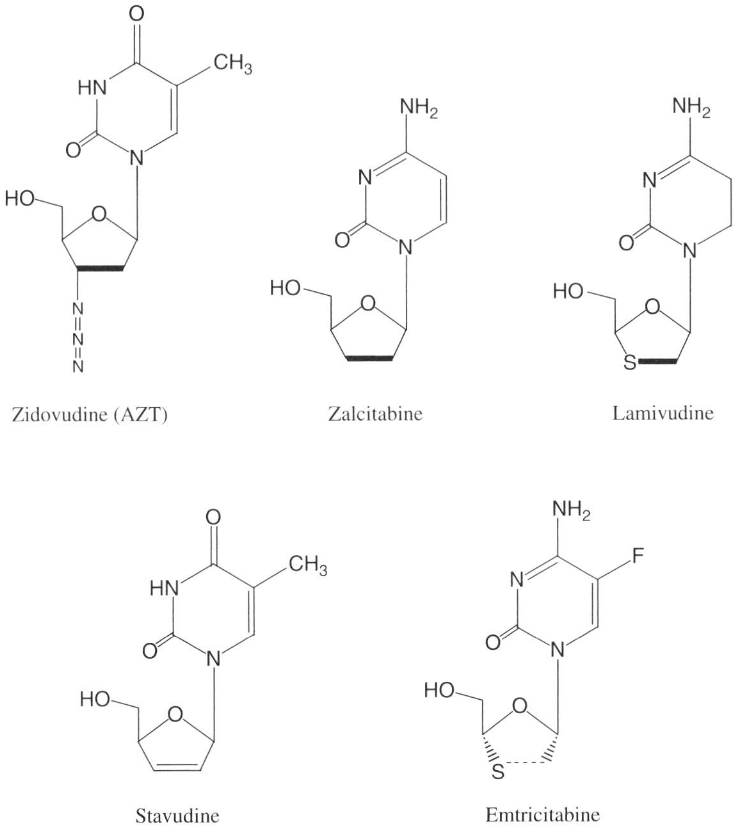

Zidovudine (AZT)   Zalcitabine   Lamivudine

Stavudine   Emtricitabine

Figure 5   Pyrimidine reverse transcriptase inhibitors.

reverse transcriptase stops that process; this proves to be lethal to the infected cell. Three of the newer drugs in this class, zalcitabine,[11] lamivudine,[12] and stavudine,[13] which retain the pyrimidine ring of AZT, similarly rely on changes in the five-membered sugar for their effectiveness. The other drug, emtricitabine,[14] includes a fluorine atom in the pyrimidine ring as well as the modified sugar.

These pyrimidine antiviral agents can be viewed as false substrates for processes that involve the corresponding DNA building blocks thymidine (T) and cytidine (C). The recognition sites on viral reverse transcriptase are apparently sufficiently degenerate so that they will also accept modified fused ring purines. Viral reverse transcriptase thus incorporates those compounds at locations intended for adenosine (A) or guanosine (G). The reverse transcriptase inhibitors didanosine[15] and abacavir[16] incorporate structural changes on the sugar that have proven useful in the pyrimidine series (Fig. 6). Adefovir is interestingly based on the open sugar series of anti-herpes agents. A further lethal modification depends on the fact that the first step in incorporation of any nucleoside, as noted previously, involves step-wise addition of phosphate groups. The presence of a phosphate-like function on

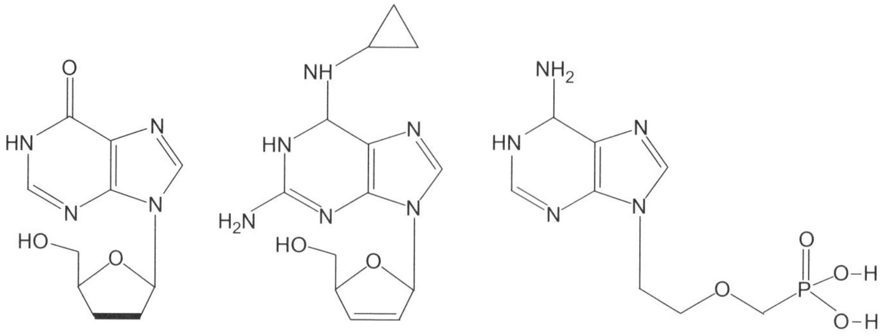

Didanosine   Abacavir   Adefovir

Figure 6   Purine reverse transcriptase inhibitors.

adefovir[17] may hasten its use by the viral enzyme. This group, as well as changes in the rest of the molecule, bring the process to a halt once incorporated, killing the infected cell.

The early promise of modified nucleosides as compounds to treat HIV infection did not halt the search for candidates from other chemical classes. Problems inherent in the nucleoside included the relatively high cost of starting materials, some drug-related side effects, as well as the observation that prolonged use led to the development of drug-resistant strains. By the middle 1980s, laboratories in both the National Cancer Institute[18] and pharmaceutical companies had in place high-capacity, semi-automated screens for identifying compounds that protected cells in tissue culture from the effects of HIV. A number of compounds were uncovered by these tests that bore no structural relation to the nucleoside (Fig. 7). In spite of their chemically very different structures, nevirapine,[19] delaviridine,[20] and efavirenz[21] all act by a very similar mechanism. Each of these molecules binds directly to a site on the reverse transcriptase enzyme given the name "non nucleoside reverse transcriptase inhibitor (NNRTI) binding pocket." It has been shown that the binding interaction of the drugs with this site leads to distortions of the three-dimensional shape of the enzyme; this change in configuration impedes enzyme function.[22] The relatively low toxicity of the compounds comes from the fact that they have no affinity for corresponding enzymes involves in DNA replication of normal cells. Although effective in vitro, the NNRTIs elicited the evolution of resistant strains even more quickly than the nucleoside transcriptase inhibitors. They are, as a consequence, almost always administered as part of a multidrug cocktail.

The protein coat that covers the virion DNA or RNA strand plays a very important role in the agents' propagation. It is that coat that first provides protection for the genome from environmental damage in its time between hosts. Some plant virions for example remain infective for years. The protein coat also provides the structures and biochemical mechanisms required for cell entry. Many viruses such as HIV have special recognition sites on the protein coat for surface features on the host immune cell white cells that are their target. Synthesis of the protein coat is thus an essential

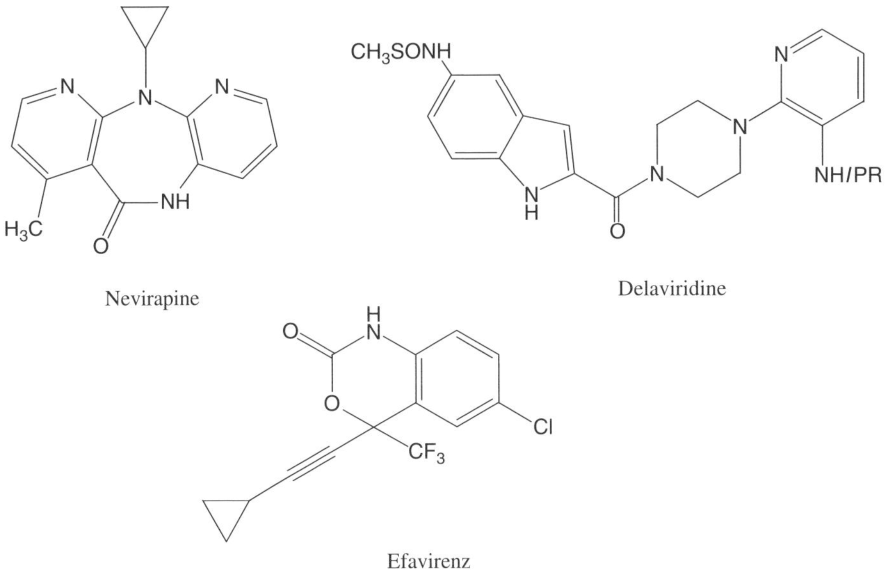

Figure 7 Non-nucleoside transcriptase inhibitors.

feature of virion survival and replication. The protein that will make up the coat is first elaborated as a much larger molecule. The HIV virion relies on a protease to cleave the larger protein to give that portion that will be used for the virus coat. Considerable unrelated search was under way in the late 1980s aimed at developing inhibitors of the peptide-cleaving enzyme renin as antihypertensive agents (see Chapter 3). It became apparent from this work that the site cleaved by renin on angiotensinogen was analogous to that which led to the HIV protein coat. The large amount of data that had by then been accumulated on renin inhibitors thus significantly advanced the discovery of small-molecule HIV protease inhibitors. Some nine protease inhibitors have been approved by the FDA for use in treating HIV patients as of this writing:

- Amprenavir
- Atazanavir
- Fosamprenavir
- Indinavir
- Lopinavir
- Nelfinavir
- Ritinovir
- Saquinavir
- Tiprinavir

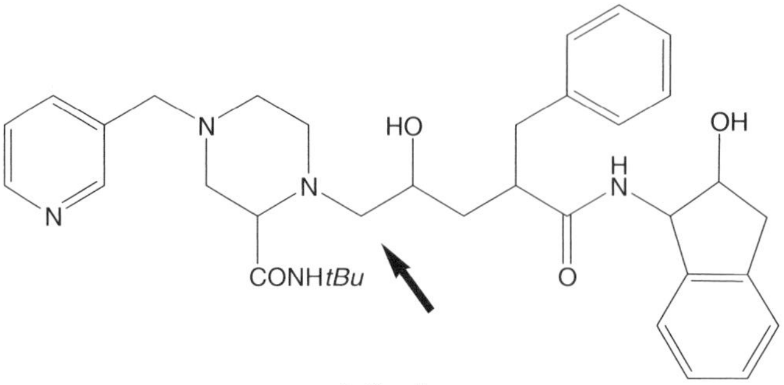

Indinavir

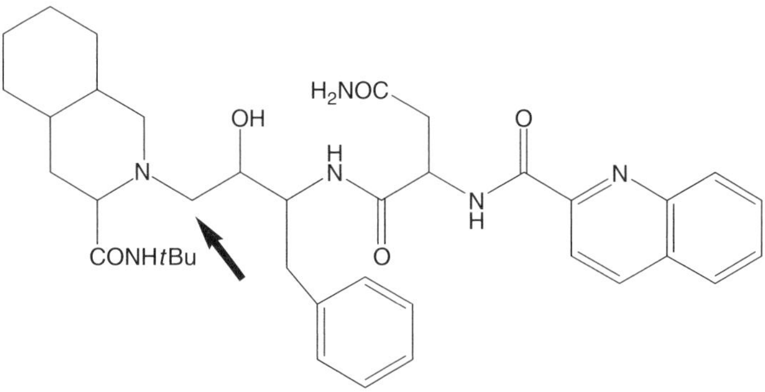

Saquinavir

Figure 8   Representative protease inhibitors.

The complex structures of these protease inhibitors and the fact that all are produced as single mirror image forms (diastereoisomers) contributes to the high cost of these very effective drugs. The chemical structures of each of these drugs, only two of which, indinavir[23] and saquinavir,[24] are illustrated (Fig. 8) in the interest of saving space, sufficiently resemble the stretch of the protein that is to be cut by the enzyme to be taken up by the protease enzyme. They also include a site, indicated by the bold arrow, designed to mimic the bond to be cleaved. That specific sequence, however, lacks the cleavable amide function. The now-bound enzyme– inhibitor complex in effect stalls and inactivates the protease. This then effectively stops the virion-coating process.[25]

In order to enter the immune system white cell that is the locus of HIV infection, the virion needs to first fuse its outer coat with receptors on the surface of the host cell. Elucidation of the detailed biochemistry of this process has led to the development of a drug that specifically inhibits the process. The compound in question, enfurvirtide,[26] comprises a synthetic 36 amino acid polypeptide. The structure of this agent partly mimics the sequence of that by which the virion binds with the host cell. The drug thus occupies the site, in effect closing off that route of attack.

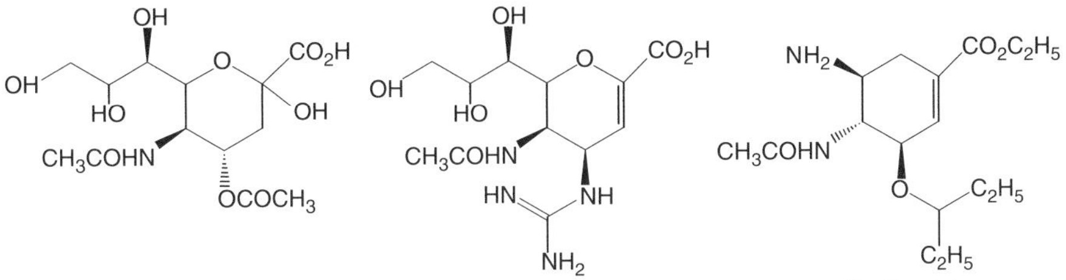

Figure 9   Neuraminidase inhibitors.

Enfurvirtide is of necessity administered by injection, because such a large peptide would not be absorbed by oral administration.

The potential for a pandemic bird influenza has focused new attention on this viral disease. One need only cast back to the period at the end of the Great War to imagine the dire circumstances that can result from an outbreak of bird flu derived disease. Up until a few years ago, amantidine and the closely related rimantidine were the only agents that showed even partial activity against the influenza virus. The much more effective recently approved drug, Tamiflu®, is the outcome of a research project that started with detailed information on the life history of influenza virus infected host cells. Formation of new influenza virions, after the DNA has been replicated, starts with the formation of a bud on the surface of the cell. The bud is at this point covered and held in place by a complex polysaccharide called sialic acid. In the normal course of events the enzyme neuraminidase is called into action to break open the sialic acid cover so as to complete bud formation and release. The first successful neuraminidase inhibitor (Fig. 9) was zanamivir, developed at Hoffman–La Roche.[27] The chemical structure of this drug incorporates many of the features of the enzyme. It thus competes for binding sites and in effect prevents completion of the budding process and release of new virions. The synthesis of this drug interestingly used for a starting material neuraminic acid, a component of the enzyme itself. The more recent and structurally less complex analog, oseltamivir[28] (Tamiflu®) has much the same activity as its predecessor.[29]

REFERENCES

1. K. Gerzon, E. V. Krumkalus, R. L. Brindle, F. Marshall and M. A. Boot, *J. Med. Chem.*, 6, 760 (1963).
2. P. E. Aldrich, E. C. Hermann, W. E. Meier, M. Paulshock, W. E. Prichard, J. A. Snyder and J. C. Watts, *J. Med. Chem.*, 14 (1971).
3. A. G. Bukrinskaya, N. K. Vorkunova and N. L. Pushkarskaya, *J. Gen. Virol.*, 60, 61 (1982).
4. J. T. Witkowski, R. K. Robins and R. W. Sidwell, *J. Med. Chem.*, 15, 1150 (1972).
5. D. E. Bergstrom and J. L. Roth, *J. Am. Chem. Soc.*, 98, 1857 (1976).
6. A. S. Jones, G. Verhelst and R. T. Walker, *Tetrahedron*, 45, 4415 (1989).
7. H. J. Schaeffer, L. Beauchamp, P. de Miranda, G. B. Elion and D. J. Bauer, *Nature*, 272, 583 (1978).

8. J. P. H. Verheyden, in D. Lednicer, Ed., *Chronicles of Drug Discovery*, Vol. 3, ACS Books, Washington, DC, 1993, p. 299.

9. B. M. Choudary, G. R. Green, P. M. Kincey, M. Paratt, J. R. M. Dales, G. P. Johnson, S. O'Donnell, D. W. Tudor and N. Woods, *Nucleosides Nucleotides*, 15, 1981 (1996).

10. J. P. Horwitz et al., *J. Org. Chem.*, 29, 2076 (1964).

11. J. P. Horwitz, J. Chua, M. Wolf and J. T. Donatti, *J. Org. Chem.*, 32, 817 (1967).

12. R. Storen, I. R. Clemmens, B. Lamont, S. A. Noble, C. Williamson and B. Belleau, *Nucleosides Nucleotides*, 12, 225 (1992).

13. J. W. Beach, H. O. Kim, N. Satynariana, I. Qamrul, K. A. Soon, J. R. Babu and C. K. Chung, *J. Org. Chem.*, 57, 3887 (1992).

14. L. S. Jeong, R. F. Schinazi, J. W. Beach, H. O. Kim, S. Nampalli, K. Shanmuganathan, A. J. Alves, A. McMillan, C. K. Chu and R. Mathis, *J. Med. Chem.*, 36, 181 (1993).

15. A. Holy, H. Dvorakova J. Jindrich, M. Masojidkova, M. Budesinsky, J. Balzarini, G. Andrei and E. De Clercq, *J. Med. Chem.*, 39, 4073 (1996).

16. S. M. Daluge, S. S. Good, M. B. Faletto, W. H. Miller, M. H. St Clair, L. R. Boone, M. Tisdale, N. R. Parry, J. E. Reardon, R. E. Dornsife, D. R. Averett and T. A. Krenitsky, *Antimicrob. Agents Chemother.*, 41, 1082 (1997).

17. J. E. Starett, D. R. Tortolani, J. Russell, M. J. M. Hitchcock, V. Whiterock, J. C. Martin and M. M. Mansury, *J. Med. Chem.*, 37, 1857 (1994).

18. For a description of the NCI screen see D. Lednicer and K. M. Snader in *Economic and Medicinal Plant Research*, Vol. 5, Academic Press, San Diego, CA, 1991, p. 5.

19. K. D. Hargrave, J. R. Proudfoot, K. G. Grosinger, E. Cullen, S. R. Capada, J. R. Patel, V. U. Fuchs et al., *J. Med. Chem.*, 34, 2231 (1991).

20. D. L. Romero, R. A. Morge, M. J. Genin, C. Biles, M. Busso, L. Resnik, I. W. Althaus, F. Reussser, R. C. Thomas and W. G. Tapley, *J. Med. Chem.*, 36, 1505 (1995).

21. M. Patel, R. J. McHugh, B. C. Cordova, R. M. Klabe, L. T. Bacheler, S. Erickson-Viitanen and J. D. Rodgers, *Bioorg. Med. Chem. Lett.*, 11, 1943 (2001).

22. N. Sluis-Cremer, N. A. Temiz and I. Bahar, *Curr. HIV Res.*, 2, 323 (2004).

23. M. D. Askins, K. K. Eng, K. Rossen, R. M. Purick, K. M. Wells, R. P. Volante and P. J. Reider, *Tetrahedron Lett.*, 35, 673 (1994).

24. K. E. B. Parkes, D. J. Bushnell, P. H. Crackett, S. J. Dunson, A. C. Freeman, H. P. Gunn, R. A. Hopkins, R. W. Martin, J. H. Merrett, S. Redshaws, W. C. Spurden and G. J. Thomas, *J. Org. Chem.*, 59, 2656 (1994).

25. For a review see R. C. Ogden and C. W. Flexner, Eds, *Protease Inhibitors in AIDS Therapy*, Dekker, NY, 2001.

26. T. Matthews, M. Salgo, M. Greenberg, J. Chung, R. Demasi and D. Bolognesi, *Nat. Rev. Drug Discov.*, 3, 215 (2004).

27. J. Scheigetz, R. Zamboni, and M. A. Bernstein and B. Roy, *Org. Prep. Proc. Int.*, 27, 637 (1995).

28. E. J. Eisenberg, A. Bidgood and K. C. Cundy, *Antimicrob. Ag. Chemother.*, 41, 1949 (1997).

29. For a recent review of this class see A. L. Liu, Y. T. Wang and G. H. Di, *Drugs Fut.*, 30, 799 (2005).

Chapter 3

Antihypertensive Agents

Hypertension has been called the silent killer. This largely symptomless condition is one of the major risk factors in cardiovascular disease. It is by now well accepted that chronic hypertension is a major causative factor in strokes and heart attacks. Unlike many other diseases, however, chronically elevated blood pressure can in some ways be viewed as simply an exaggerated form of a normal physiological process.

Humans, in common with all mammals, are utterly dependent on the circulatory system. Individual cells, to say nothing of organs, require a constant supply of nutrients, oxygen, and various hormones, enzymes, and the other compounds that control their minute-to-minute function. They also require a system to carry away metabolic waste products. The intricate system of vessels that fulfills this function includes ones that range in size from the rather large aorta to capillaries so fine that red blood cells pass through one at a time. The force required to push the somewhat viscous carrier fluid, blood, through this network is expressed as its pressure. The critical nature of this function is reflected by the complex interconnected systems that maintain blood pressure within the range required for efficient function. The nervous system, for example, is involved at several levels: it adjusts heart rate and the force of cardiac contraction and controls resistance to flow by constricting or relaxing the vessels that terminate the arteries, the arterioles, further downstream. The chemicals, called neurotransmitters, that translate the signals from the nervous system comprise another control system. The familiar fight or flight response, which leads to a temporary increase in heart rate and blood pressure, is mediated at this level. An alarming event will, for example, cause the release of a flood of the neurotransmitter adrenaline. This will cause a sudden increase in heart rate and blood pressure associated with panic attacks. Pressure is also directly related to the amount of fluid in the circulatory system in accordance with classical laws of hydrodynamics for a closed system. The kidney and its associated hormones are intimately involved in regulating blood volume. They thus play their own role in controlling blood pressure. These various systems are interconnected by an intricate set of feedback responses that maintain pressure at an optimal level under normal conditions. These many mechanisms involved in the control of blood pressure have led to the development of a wide array of drugs for addressing the problem at correspondingly varied targets. It is interesting to note, in passing, that in the vast majority of patients, the cause of their hypertension is not traceable to malfunction of any specific part of the regulatory system. Their disease is simply consigned to the diagnosis "essential hypertension."

The discovery and development of drugs for lowering elevated blood pressure have involved interplay between serendipity and basic science. Many early drugs were discovered by screening compounds produced in the labs of organic chemists in animal models for the disease. More detailed investigations of how the drugs actually worked gave a wealth of information about the various pathways by which blood pressure is actually controlled. This knowledge, along with data from pure basic research, then led to efforts to design drugs that acted on the newly uncovered pathways. Instead of looking for compounds that simply lower blood pressure, biologists today tend to seek agents that will act on a specific pathway that, in the end, controls pressure.

Their place in nature as the only sentient creatures provides a measure of comfort to most humans. This feeling is given the lie by the fact that the most basic functions that allow us to carry on are controlled by the autonomic rather than the central nervous system, in other words, the so-called reptile brain. This network, more familiarly called the involuntary nervous system, controls almost all vital life systems. This, not surprisingly, includes control of blood pressure. As noted above, pressure is dependent on blood volume, the heart's pumping rate and force, and, finally, resistance to flow. This last factor is expressed by the diameter of the arterioles, the vessels between the arteries and the capillaries. The diameter of these blood vessels is directly controlled by smooth muscles. Signals provided by the autonomic nervous system comprise one of the main means by which the muscles cause a vessel to contract or dilate. This system in turn responds to a set of sensors, called baroreceptors, that react to blood pressure.

Transmission along nerves differs from conduction along wires in that it involves alternation between electrical and chemical signals. Transmission is also a one-way process. An impulse will thus travel down a neuron as an electrical charge until it reaches the end of that cell. The gap between that neuron and the next nerve cell, called a synaptic cleft, or often simply a synapse, is equipped with special structures at the downstream end called receptors; these in turn react to chemical signals. Arrival of a signal at the synapse will cause the release of a special chemical termed a neurotransmitter. When this crosses the gap and meets a similar structure on the next nerve fiber, it will initiate an electrical signal in that neuron that then travels to the next synapse. Structures called ganglia, located alongside the spinal column, comprise one of the first synapses met by a signal that originates in the autonomic nervous system. Early research on nerve transmission had identified the relatively simple charged molecule acetylcholine as the neurotransmitter in ganglionic synapses. Animal experiments with some even smaller charged compounds, which included the charged part of acetylcholine called a quaternary ammonium salt, showed that these could be used to block nerve transmission

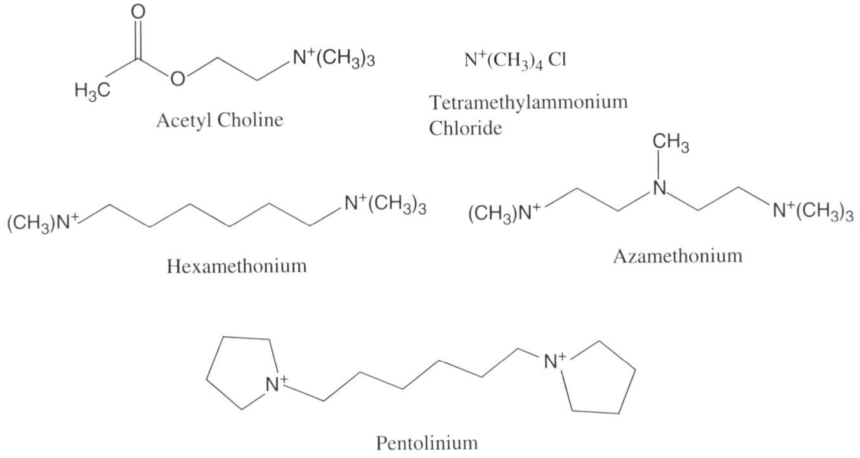

Figure 1   Ganglionic blockers.

across ganglionic synapses, and blood pressure did indeed drop when animals were treated with these simple salts. The hosts of other autonomic functions that were blocked as well ruled out the use of these for treatment of blood pressure. Taking this as a clue, scientists at May and Baker in England synthesized a series of molecules that included charged quaternary ammonium pieces. One of these, hexamethonium, was eventually found to be useful in lowering blood pressure in animals.[1,2] These results were repeated in trials in humans, where the drug gave dramatic results in lowering blood pressure in patients with severe malignant hypertension. Later controlled trials confirmed the utility of this drug as an antihypertensive agent.[3,4] Hexamethonium and related ganglionic blocking agents (Fig. 1) were unfortunately not very selective in that they also blocked a multitude of other functions. The resulting poor tolerance restricted their use to patients with malignant disease. Several other ganglionic blockers, including azamethonium and pentolium have been marketed as well. The more selective and better tolerated drugs that came along within a decade relegated these compounds to a footnote in the history of drug research.

Transmission of nerve signals further down the line involves a pair of chemically very closely related neurotransmitters. Compounds whose structures are based on these molecules provide the basis for many drugs that are still widely used for treating hypertension. Both adrenaline, also known as epinephrine, and noradrenalin, alternatively called norepinephrine, had been isolated from animal adrenal glands early in the twentieth century. Research over the next decades identified the compounds as the mediators of the so-called "fight or flight" response mentioned earlier. The flood of epinephrine released by the adrenals in response to an alarming event led to an almost immediate set of responses including accelerated heart rate and increased blood pressure. To summarize, and arguably oversimplify, years of intensive research subsequently identified these molecules as neurotransmitters in the sympathetic nervous system. It was later shown that specific receptors showed

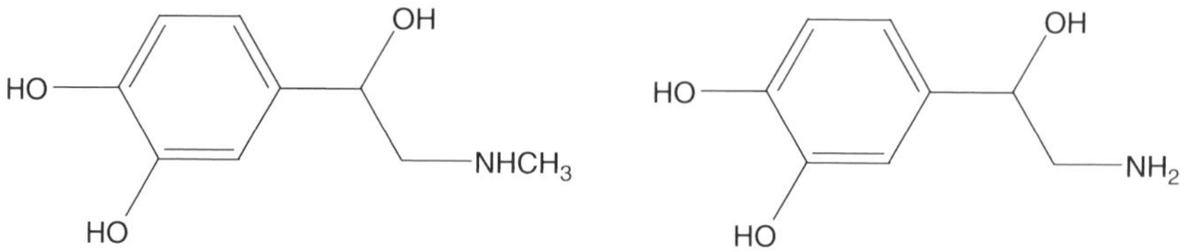

Epinephrine
Adrenalin

Norepinephrine
Noradrenalin

Figure 2   The catecholamines.

different sensitivities to these two molecules; they were thus designated as α- and β-sympathetic receptors. Epinephrine, which acts on both the α- and β-sympathetic receptors, lowers blood pressure at low doses, but has the reverse effect at higher dose. Norepinephrine on the other hand acts largely on α-sympathetic receptors to increase blood pressure. These compounds are often referred to collectively as catecholamines after their chemical structures (Fig. 2). Drugs that would antagonize norepinephrine, soon called α-blockers, should in theory lower blood pressure. It should be borne in mind that this pioneering work on neurotransmitters was carried out long before the availability of today's tools. Scientists in those days relied almost entirely on methods that involved strips of muscle tissue suspended in organ baths. They would then study the contraction or relaxation caused by adding a compound to the bath as well as that response to a putative blocker. Current methods rely on such in-vitro procedures as binding of drugs to preparation of isolated receptors preparations and various immunoassay-based tests.

One of the initial leads for new drugs for treating high blood pressure paradoxically came from a compound that had α-sympathetic activity. In the late 1930s, scientists at CIBA had found that a newly synthesized compound caused vasoconstriction.[5] This agent, naphazoline, later classed an α-agonist, proved to be a useful nasal decongestant by constricting blood vessels in nasal passages. The general structure was to prove exquisitely sensitive to small changes. Simply omitting one of the fused rings led to the creation of tolazoline. This molecule interestingly lowered blood pressure by acting as a vasodilator. It is still used today, though mostly in order to dilate arteries in conjunction with medical procedures. A somewhat more complex molecule, which also included the five-membered imidazoline ring, also blocked the action of α-sympathetic catecholamines[6] and thus led to vasodilatation and reduced peripheral resistance. Clinical trials proved this compound, phentolamine, to be a useful antihypertensive agent. In the same period, chemists at Smith, Kline & French synthesized an α-blocker that lacked the imidazoline ring.[7] This drug, phenoxybenzamine, also found use in the treatment of hypertension. This last drug reacts irreversibly with the receptors and is thus used mainly for diagnostic purposes. These and several related α-blockers (Fig. 3) found a clear advantage over the ganglionic blockers with regard to tolerability. They still, however, shared a common set of side effects. Most prominent of these are increased heart rate and fluid retention. These and other effects have been attributed

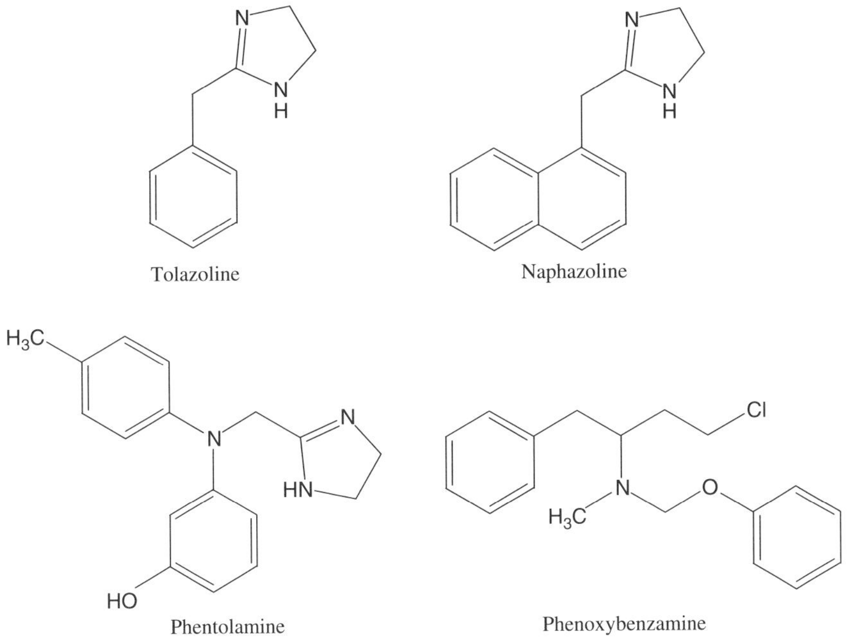

Figure 3   Alpha-adrenergic blockers.

to compensatory responses where the cardiovascular system tries to overcome the effect of lowering blood pressure. This goes along with one school of thought that holds that hypertension involves a resetting of pressure that the circulatory system perceives as normal to a higher level. This seemingly inescapable flaw led to neglect of this field for several decades. Use of these drugs is now largely restricted to the small percentage of patients whose high blood pressure is a result of phaeochromo-cytoma, a tumor of the adrenals that results in excess production of epinephrine and norepinephrine. The drugs in this case act directly against the substance that leads to the symptoms of the disease.[8,9] Patients who suffer this disease are among a small group for whom the cause of hypertension can be clearly explained.

The formalities imposed by scientific journals make candid accounts of the events that lead to the discovery of new drugs more the exception than the rule. The story behind the development of another antihypertensive agent that involves the adrenergic system comes from an account published by its discoverer, Helmuth Stähle.[10] In 1960 this scientist at the Boehringer Ingelheim firm in Germany launched a program to prepare novel nasal decongestants. The chemical structures of some of the earlier known vasoconstrictor α-agonists were used as guides for this work. The structures of the target compounds, as a result, included a functional group that is related to that present in tolazoline and also, for that matter, phento-lamine. Testing the various compounds from the program in an animal model led to the identification of one congener that was deemed worthy of further investi-gation. It is generally recognized that the human is the ultimate test organism for any drug intended for use in humans. Many fewer obstacles and much less paper work existed in going from animal to humans tests in those earlier, more innocent days. A physician in the trials group thus administered the drug, still known by its internal code number, St 155, to his secretary, who happened to have a cold. Much to his "... surprise and embarrassment the lady fell asleep for 24 hours." Further investigation of this contretemps revealed that the drug, soon renamed clonidine, was a potent antihypertensive agent. The dramatic effect was apparently caused by sedation and hypotension caused by the drug being absorbed through nasal mucosa. Elucidation of the mechanism of action uncovered a hitherto unknown pathway for controlling blood pressure. Although the drug is both a weak α-adrenergic agonist and antagonist on vascular receptors, it interacts strongly with receptors in the brain. This interaction then leads to a decrease in blood pressure. A small group of structurally related compounds that contain a guanidine function, including guanabenz and guanfacine, which share many structural features with clonidine, were found to act by the same mechanism (Fig. 4). Clonidine and the related drugs are still currently prescribed as antihypertensive agents in special cases.

Many pharmaceutical companies maintained animal screening tests in the search for widely effective and well-tolerated antihypertensive agents. These assays employed various animal systems, ranging from rats with normal blood pressure to an inbred strain that had elevated blood pressure. In the hope of uncovering novel structural leads, these screens were used to test not only chemicals designed by their chemists as antihypertensive agents, but also a selection of compounds designed for other targets. That the compound that was later named prazocin bore no structural relation to then-known antihypertensive drugs suggests that the series[11,12] may well have been discovered in the Pfizer labs in a random screen. Classical pharmacology at first suggested that this antihypertensive was just another α-blocker. It might well have fallen by the wayside because of the well-known drawbacks of this class of drugs. However, there would seem to have been enough small differences in the pharmacology of these compounds from known α-blockers to encourage further research.[13] Recall that α-adrenergic receptors are a subclass of the synapses that bind catecholamines. Detailed pharmacology prompted by the discovery of prazocin led to the discovery that α-adrenergic receptors were divided into two categories. Whereas previous blockers bound to both

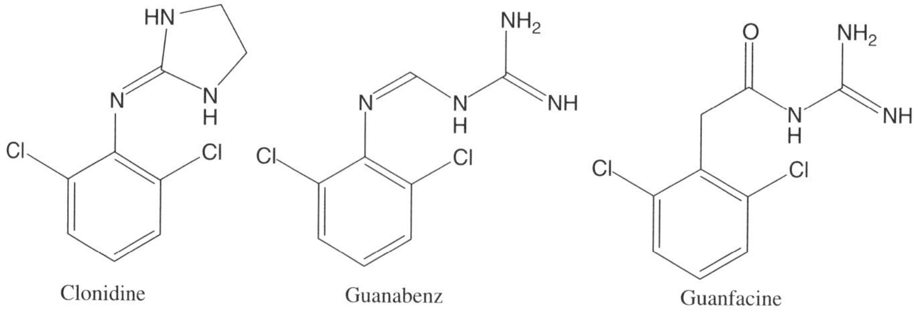

Clonidine	Guanabenz	Guanfacine

Figure 4   Clonidine and other α-2 blockers.

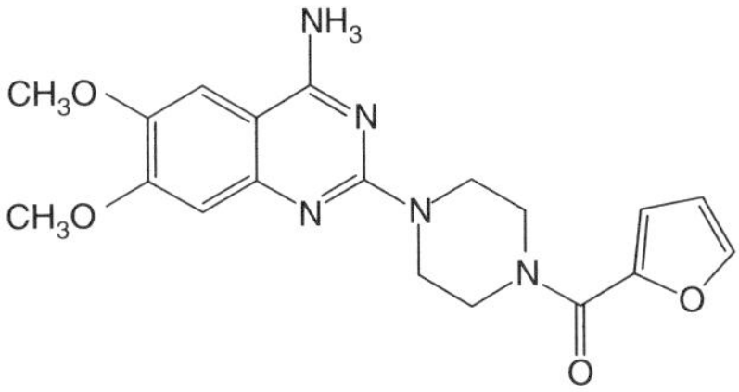

Figure 5   Alpha-1 adrenergic antagonists.

subclasses, prazocin interacted specifically with the newly identified subclass, dubbed α^1-adrenergic receptors (Fig. 5).[14] This scenario, where the need to investigate the biological properties of a promising new chemical entity uncovered novel pharmacology, was to be repeated more than once in subsequent drug development in this and other fields. The effectiveness and relatively good tolerance of prazocin was established not long after.[15,16] Some half dozen related compounds have been granted nonproprietary names, of which three are on the market in the United States. All these compounds shared the same fused ring nitrogen-containing fragment. Widespread use in the clinic led to the finding that they provided relief of symptoms from enlarged prostates. This is now an important indication for several of the α^1-blockers after the finding was confirmed in clinical trials.[17] Relief of the symptoms from an enlarged prostate is in fact one of the principal indications for one of these agents, trimazocin. It is believed that the drug owes its effect to binding to α^1-receptors in the prostate.

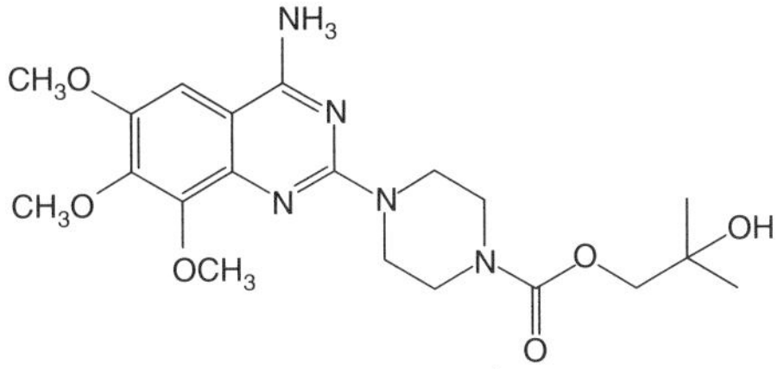

Recall that activation of α-adrenergic receptors leads to vasoconstriction, hence the effort to focus antihypertensive drugs on blocking that receptor. Interaction of catecholamines with β-adrenergic receptors on the other hand results in general relaxation of involuntary muscle. Epinephrine itself, and its simple derivative isoproterenol, was used quite early on to treat asthma because of its relaxant action on bronchi. The relatively short duration of action of these agents was traced to quick metabolic destruction. The first-generation drug was isoproterenol (Fig. 6), epinephrine in which the methyl group on nitrogen is replaced by isopropyl ($-CH(CH_3)_2$) in order to circumvent an enzyme that removes methyl groups on nitrogen. In the search for a yet longer acting drug, in 1958, scientists at Lilly prepared a compound in which the two hydroxyl groups in isoproterenol were replaced

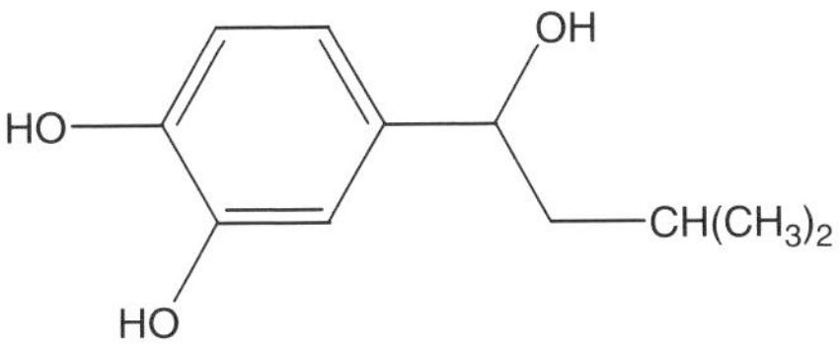

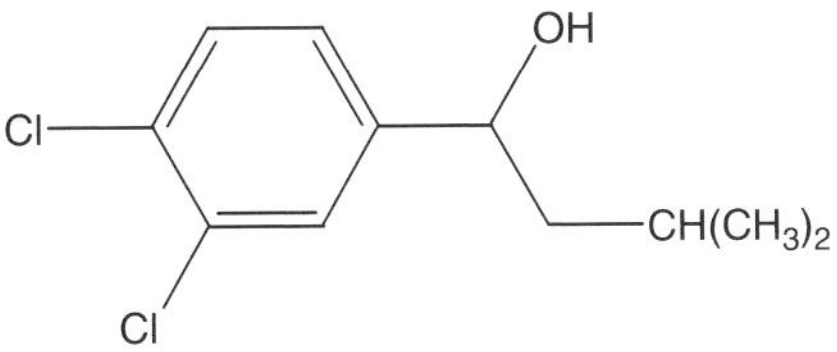

Figure 6   Isoproterenol and DCl.

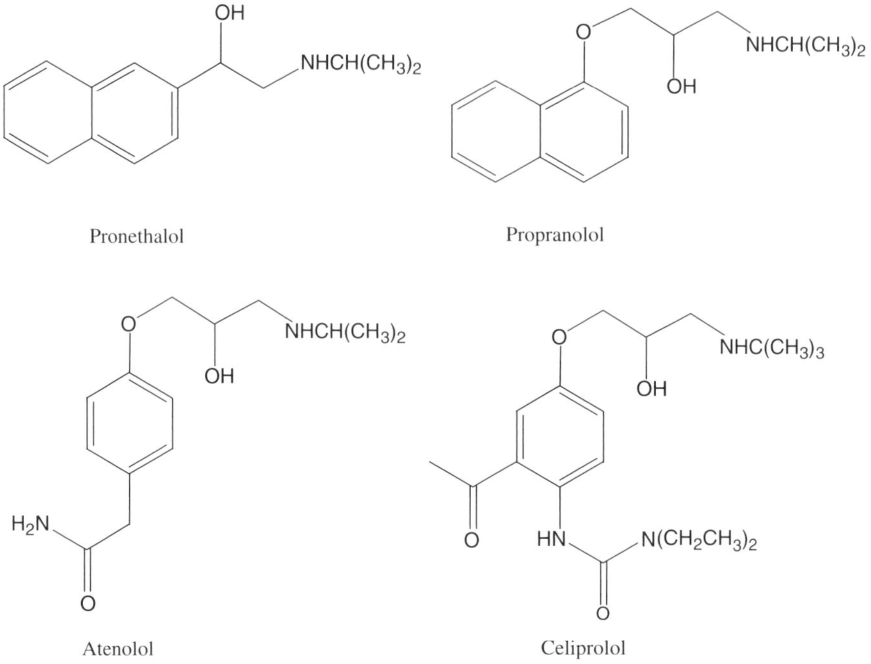

Pronethalol

Propranolol

Atenolol

Celiprolol

Figure 7   Some beta-blockers.

by chlorine. The product, dichloroisoproterenol, later better known as DCl, surprisingly blocked rather than added to the effect of epinephrine on receptors (Fig. 6). At the same time, DCl paradoxically lowered blood pressure in selected animal models for hypertension when logic would indicate that it should increase pressure by causing vasoconstriction.

Detailed pharmacology then showed that the compound in fact blocked the effect of epinephrine in various experiments. This compound was the first of what was to be a long series of β-blockers (Fig. 7). DCl itself, however, had too many drawbacks to be developed as a drug. Taking this as a clue for further work, scientists at ICI in the UK, under the leadership of future Nobelist (and Sir) James Black, launched a program to develop compounds based on that finding. This culminated in the preparation of pronethalol.[18] The compound, which has an additional fused ring replacing the two chlorine atoms, was taken to the clinic to test its effects on several cardiac diseases, including arrhythmias, angina, and hypertension. The activity of this early β-blocker as an antihypertensive agent in humans was soon confirmed.[19] Pronethalol's place in pharmacopeias was to prove short-lived, as the compound was found to cause tumors in animal models. Its replacement, propranolol, was however waiting in the wings at ICI and was quickly introduced into clinical practice. Although the main indication is hypertension, the drug finds considerable use as an antianginal agent. Propranolol is quite effective and generally free of disabling side effects. The success of this drug led to intensive research in many pharmaceutical laboratories. This resulted in the preparation of a veritable flood of β-blockers. Close to two dozen of these drugs have been assigned nonproprietary names. Fourteen are approved for use in the United States.[20] The work in this series revealed that the side chain ($-OCH_2CHOHCH_2$ NHCX) is an absolute requirement for activity. There is considerable freedom as to the chemical structure of the fragment that holds the side chain as shown by two compounds selected almost at random, atenolol and celiprolol. An apocryphal story has it that a British medical journal once published a spoof on the properties of a newly introduced drug named Ololol. The exact mode of action of β-blockers is still not completely understood. The fact that they lower cardiac output by decreasing heart rate and force of contraction may play a large role in the blood pressure lowering effect. This same property has led to informal use of these drugs to combat the symptoms of stage fright. More than one performer is said to have taken a tablet of one of these drugs before a performance.

Beta-blockers have, perhaps unexpectedly, been found useful for treatment of glaucoma (Fig. 8). This disease includes damage to the visual field caused by elevated pressure of the fluid within the eye. It was found not too long after the introduction of propranolol that the agent lowered intraocular pressure.[21] Beta-blockers that more effectively lowered this pressure were found as additional compounds in this class became available. The drugs are believed to work by decreasing aqueous humor formation.[22] The drugs are administered locally to the eye as drops in order to avoid the systemic effects. Treatment of glaucoma is one of the more important applications of these drugs.

Fluid volume is one of the important parameters in determining pressure within a closed system. Simple analogy would suggest that the volume of blood in the circulatory system should play an important role in hypertension. The discovery of the nonmercurial diuretic drugs traces back to the observation that some sulfonamide antibacterials caused excess diuresis. The availability of very effective and safe

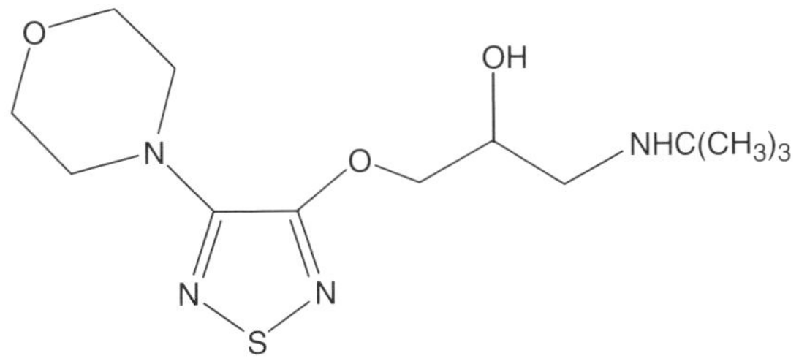

Timolol

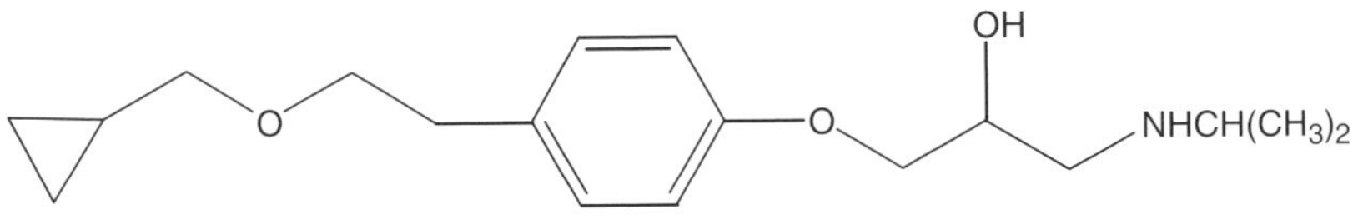

Betaxolol

Figure 8   Beta-blocker for glaucoma.

diuretic drugs such as chloraminophenamide[23] in the late 1950s provided a means for safely reducing blood volume. Indeed, it was found that this drug did bring about a moderate lowering of blood pressure in patients.[24,25] The following years saw the introduction of a large number of diuretic drugs based on the same structural model. Forming a ring between two adjacent substituents on the molecule led to hydrochlorothiazide, a diuretic with increased potency. Chlorthalidone, another widely used compound, shows that a single sulfonamide group (SO_2NH_2) is sufficient for activity. The mechanism by which these drugs lower blood pressure is not known in detail. It likely goes beyond simple volume reduction; this does drop initially but returns to pretreatment levels within a short time.[26] The initial drop in peripheral flow resistance does however persist. This may also involve a drug effect on serum sodium levels.[27] These so-called thiazides (Fig. 9) today comprise first-line therapy for patients with mild to moderately elevated high blood pressure. They are reasonably affordable, because patents on these have long ago expired. One of the most often prescribed, hydrochlorothiazide, is close to being a tonnage chemical.

A seemingly minor side effect of this class of drugs has led to the odd circumstance that some products that contain potassium salts as their only active ingredient are subject to the same FDA scrutiny and approval as a very powerful drug. This originates with the fact that thiazide diuretics tend to cause excessive loss of potassium. Urine does not of course consist of pure water. Various ions are excreted in addition to waste products. In the absence of drug the relative levels of sodium

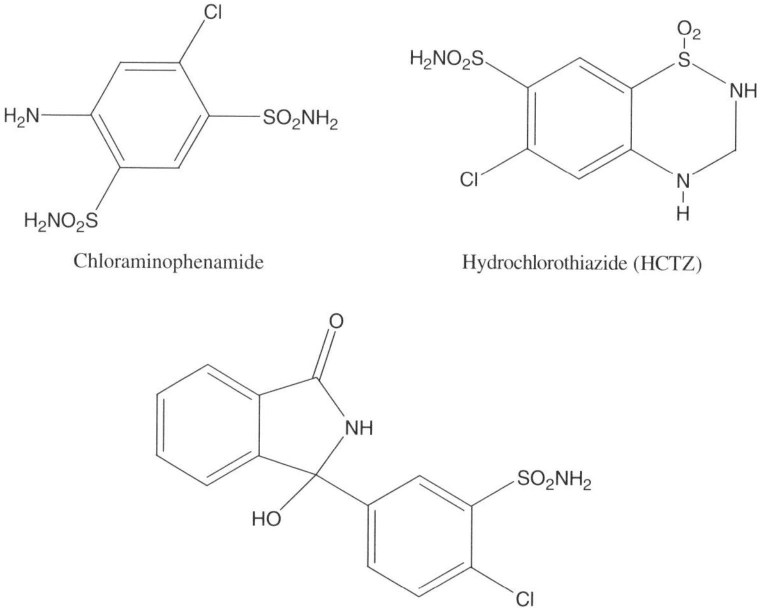

Chloraminophenamide   Hydrochlorothiazide (HCTZ)

Chlorthalidone

Figure 9 Thiazide diuretics.

and potassium ions closely reflect those in blood. Excessively low levels of serum potassium can cause a host of problems. Patients on chronic thiazide regimes are consequently often counseled to supplement their potassium intake. Those who have not tasted potassium chloride might suggest simply sprinkling that salt on food instead of sodium chloride. A single taste will quickly dispel that notion. Early supplements of potassium consisted of compressed potassium salt tablets. Swallowing those quickly, it was felt, would avoid the bad taste. Scientists at Lilly went one step further by developing tablets with a coating that would not dissolve until after they had passed beyond the mouth. This so-called enteric coating would avoid all contact with that part of the GI tract capable of taste perception. A short time after the product was placed on the market in the middle 1960s, there ensued a series of reports of severe intestinal ulcers and a few cases of actual perforation among patients who had ingested the enteric tablets. The wounds were later traced to high local concentrations of potassium that resulted when the coating dissolved away. After the product was recalled, the FDA issued a regulation mandating that all controlled or delayed release forms of potassium had to henceforth undergo the same clinical trials, and the same regulatory and review process, as a drug with true pharmacological activity.

The development of drugs discussed thus far has relied at least in part on some biological rationale. The initial lead that led to prazocin did, as noted earlier, in all likelihood come out of a random screening program. Pharmacologists did, however, soon develop data that allowed them to include the drug in the broad class of agents that act on the adrenergic system. A small number of compounds that came out of true random screening have been categorized as direct acting vasodilators (Fig. 10). Test data suggested that they act directly on vascular smooth muscle to dilate arterioles without invoking the adrenergic system. These compounds were actually discovered not too long after hexomethonium. The synthesis of the first of these, by scientists at CIBA, in fact dates back to 1950.[28] Clinical trials on this drug, hydralazine, were published not too long after.[29] The drug was initially used for treating high blood pressure, an application that declined as better tolerated compounds became available. An important current use of the drug is the treatment of congestive heart failure. The compound helps the weak heart pump blood by dilating peripheral blood vessels and thus reducing resistance to flow. One of the new start-up biomedical concerns, Nitromed, recently designed a combination

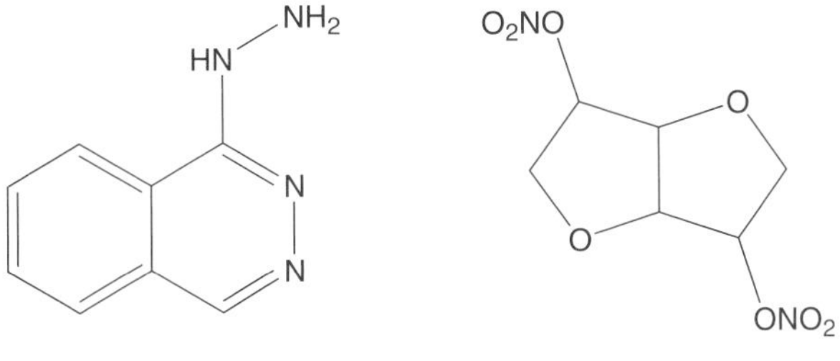

Hydralazine   Isosorbide dinitrate

Figure 10   Hydralazine and isosorbide dinitrate.

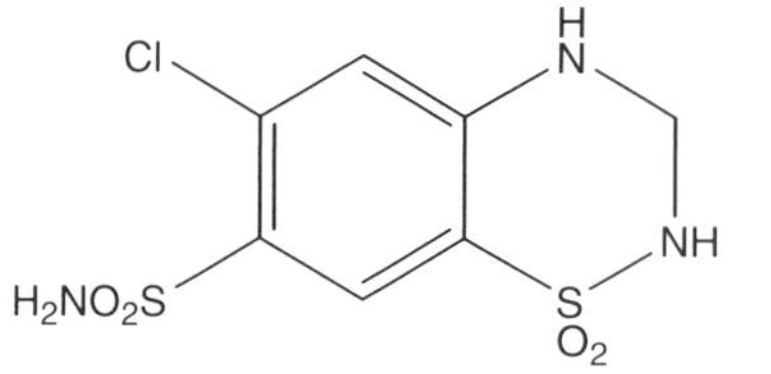

Figure 11 Diazoxide and HCTZ.

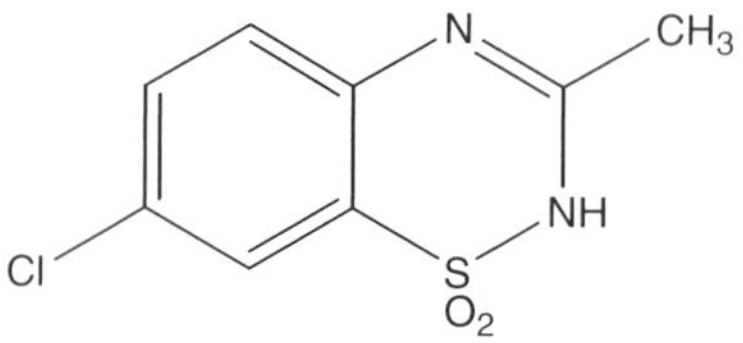

that takes advantage of that activity by testing a combination tablet that includes hydralazine and the venerable antianginal compound isosorbide dinitrate, a close relative of nitroglycerin. Statistical analysis of data from clinical trials of the combination showed very equivocal results; the data did however show unexpected statistically significant efficacy in a subgroup formed of African-Americans.[30] FDA approval of the combination drug BiDil® for use in this restricted population has engendered considerable discussion in both the medical and lay press.

The close resemblance of the chemical structure of diazoxide (Fig. 11) to that of thiazide diuretics suggests that this compound may have been originally synthesized in the laboratories of Schering (U.S.) with the latter target in mind. Biological screening revealed that the molecule is a very potent vasodilator,[31] devoid of diuretic activity. This very powerful vasodilator is most often used for treatment of hypertensive emergencies. Recent work has demonstrated that the drug relaxes smooth muscle by modulating the movement of potassium ions involved in muscle action.

The discovery and subsequent development of another vasodilator, minoxidil, provides an almost textbook example of serendipity (Fig. 12). In the early 1960s, scientists at Upjohn launched a screen aimed at finding compounds that lowered blood pressure in rats. The wide variety of chemicals put through this screen was not restricted to entities synthesized by the company chemists. Instead, they also included compounds obtained by browsing through commercial catalogues. One of the early hits, di-allylmelamine, is a tonnage chemical. This was, as the name suggests, closely related to one of the two components of melamine resins. The synthesis program that was started to exploit this lead found that the molecule could be simplified and that one of the ring nitrogen atoms could be omitted. Biologists were, however, puzzled by the lag time between administration of the simplified compound and the onset of action. Detailed biochemical experiments revealed that passage through the liver added an oxygen atom to the molecule.[32] The antihypertensive activity of the first administered compound was in fact due to this transformation product. Scientists then started a program to produce modifications of compounds related to the initial lead that now included the extra oxygen. A closely related analogue of the metabolite was deemed more suitable for further development.[33] This compound was eventually granted the nonproprietary name minoxidil. The suspicion, however, lurked in the minds of some Upjohn scientists that this oxide was still not the molecule responsible for the drug's antihypertensive

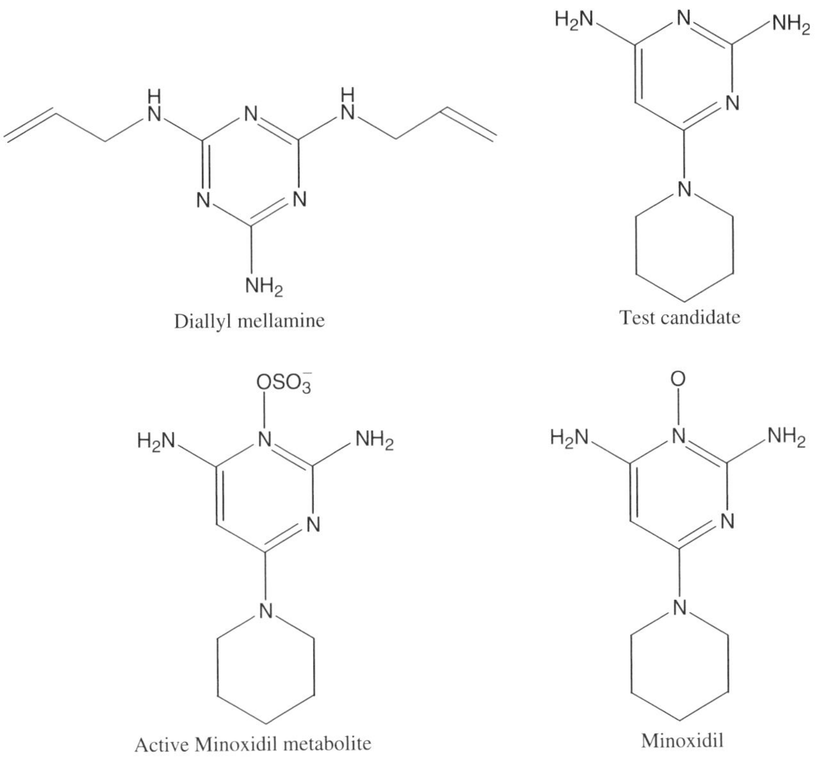

Diallyl mellamine

Test candidate

Active Minoxidil metabolite

Minoxidil

Figure 12   The minoxidil story.

activity. Further investigation many years later was rewarded with the discovery of the compound actually responsible for activity. This was an unusual O-sulfate that resulted from an additional metabolic step.[34] This last compound was too difficult to prepare and too unstable to replace minoxidil. The activity of the direct acting vaso-dilator, minoxidil, held up in clinical trials.[35] Its powerful action and side effects have restricted use of the drug to patients whose elevated blood pressure does not respond to other drugs. In a compensatory response, reminiscent of that of α-blockers, minoxidil causes an increase in heart rate and fluid retention. The drug is therefore usually given along with a β-blocker to control heart rate and a diuretic to fight water retention.

As the use of minoxidil in medical practice spread, reports of a very unusual side effect started filtering back to Kalamazoo. Patients who had been taking the drug for any length of time starting developing hair growth in unexpected places and quantity. This quickly raised the possibility that this drug might also provide a treatment for hair loss. Memory still presents the image of a colleague at Upjohn who showed up at meetings wearing a drug-covered gauze pad on his bald head. More formal and respectable trials confirmed this activity. It was,

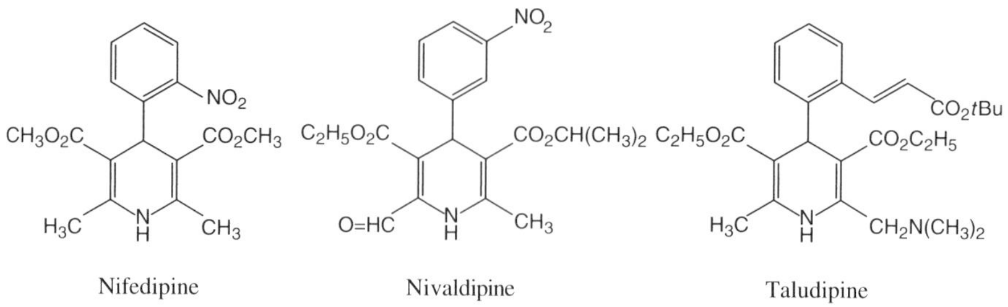

Figure 13   Dihydropyridine calcium channel blockers.

however, necessary to demonstrate that topically applied doses of this extremely potent drug did not get into the bloodstream.[36]

The discovery of the first member of yet another class of direct-acting antihypertensive agents is probably also the outcome of a screening program. Scientists at Bayer reported in 1970 on the antihypertensive activity of a class of compounds known as dihydropyridines (Fig. 13).[37] Detailed pharmacological investigation of these compounds suggested that they would be useful in treating both angina and high blood pressure. The antihypertensive activity of the compound chosen for further development, nifedipine, was confirmed in the clinic.[38] Subsequent investigation revealed that this drug acted as a calcium channel blocker. The flow of calcium ions across the membrane of a muscle cell plays a central role in the contraction of that structure. The force and duration of action of muscle will thus depend on that flow. Excess translation of calcium ions into the muscles that line blood vessels leads to vasoconstriction and resistance to the passage of blood. The same excess in cardiac muscle leads to increased force of contraction. These effects will increase blood pressure and contribute to the anginal episodes. The calcium channel blockers act by preventing excess inflow of these ions into the cells. This in effect leads to milder contractions, which, in turn, results in a decrease in the force of contraction of cardiac muscle and decrease in resistance to flow at the arterioles. The net result is a decrease in blood pressure. The effect on heart muscle is particularly useful in treating angina. Work in other laboratories led to the development of about ten other dihydropiridines with nonproprietary names, of which five are available in the United States. All these compounds include at their core the six-membered nitrogen-containing ring. Considerable latitude exists as to the appendage that is attached to this central element, as illustrated by nivaldipine and taludipine.

Elucidation of the mechanism of action of the calcium channel blockers helped explain the mode of action of some venerable drugs such as verapamil. This drug, whose structure differs markedly from the dihydropyridines, had been used to treat cardiac disease since the 1960s although its mode of action was quite speculative. Several more recent calcium channel blockers whose structures also differ from that group, for example diltiazem, are used primarily as antianginal agents (Fig. 14).

Kidneys play a very central role in regulation of blood pressure. These organs control blood volume and ensure the conservation of the various ions such as

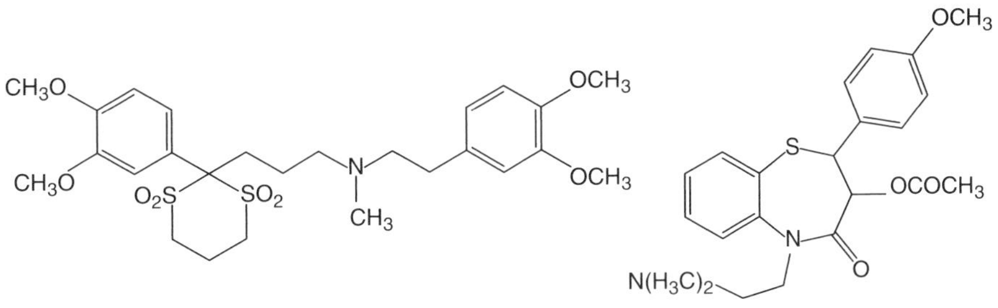

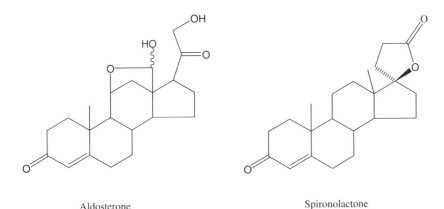

Verapamil   Diltiazem

Figure 14   Antianginal drugs.

sodium, potassium, and calcium necessary for normal muscle function. One of the hormones that controls kidney function is an unusual steroid called aldosterone, secreted by the adrenal cortex. This compound acts to conserve sodium ions and consequently promotes water retention. This was the last of the steroid hormones to have its chemical structure defined. Scientists were intrigued by the antidiuretic activity of the so-called "amorphous fraction" that remained after all other steroids had been pulled out of adrenal extracts. Its structure was finally determined in 1954, when advances in chemistry made it possible. In the early 1950s, chemists at G.D. Searle launched an intensive program on steroid synthesis that culminated in the development of the first oral contraceptive. Their biological colleagues at that time maintained a set of standard screens that included one that would detect diuretic agents. One of the steroids from the chemical program turned up as an active diuretic.[39] The compound chosen for clinical trials, spironolactone, was found to antagonize aldosterone and its sodium- and water-retaining activity (Fig. 15). There are surprisingly few similarities between the structures beyond the steroid nucleus between the endogenous compound and its antagonist. Spironolactone, like its thiazide counterparts, also lowered blood pressure. The compound

Aldosterone   Spironolactone

Figure 15   Aldosterone and spironolactone.

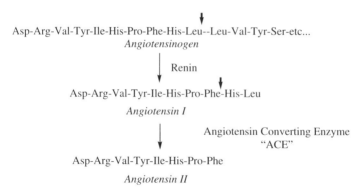

Figure 16   The renin–angiotensin system.

did not, however, unlike the latter, result in excess potassium loss.[40] This drug and several related later entries have found only minor use in the treatment of hypertension, possibly as a result of their high cost compared to thiazides.

Kidneys further impact on blood pressure by releasing a hormone that acts directly on blood vessels so as to control resistance to flow. This effect is mediated largely through the so-called renin–angiotensin system. The process starts with a large peptide, angiotensinogen, which has little if any activity in its own right. In response to an appropriate signal, another kidney peptide enzyme, renin, splits out a smaller, ten-amino-acid piece from angiotensinogen called angiotensin I. Yet another enzyme, called appropriately angiotensin-converting enzyme, (ACE) then removes two amino acids from the end to give the potent vasoconstrictor angiotensin II. Release of this last substance of course raises blood pressure. The sequence is shown in outline form in Figure 16 (amino acids are designated by standard abbreviations).

A group at Squibb investigated a series of peptides isolated from a snake venom that inhibited the angiotensin-converting enzyme (Fig. 17).[41] They then synthesized a series of small eight-amino-acid peptides whose structure was based on the most active constituent from the venom. One of these analogs, teprotide, which comprised nine amino acids, lowered blood pressure in animal models for hypertension. It also did so, most notably, in human volunteers. The fact that peptides are often poorly absorbed and quickly destroyed in the body led the group to seek a synthetic small molecule that had the same activity. They had accumulated sufficient data on how changes in the chemical structure of the analogues of teprotide affected ACE-inhibiting activity to construct a working model of the site of action. They then synthesized a series of analogs based on that model. The compound chosen for clinical workup, captopril, was the first antihypertensive ACE inhibitor to be approved for use in humans. This drug is also one of the first cases where knowledge of the active site of drug action was successfully applied in the design of a drug. A number of analogs from competing laboratories soon joined captopril on the market. These later drugs all avoided the use of the sulfur-containing group found in captopril, an entity that had been associated with some unpleasant side effects.

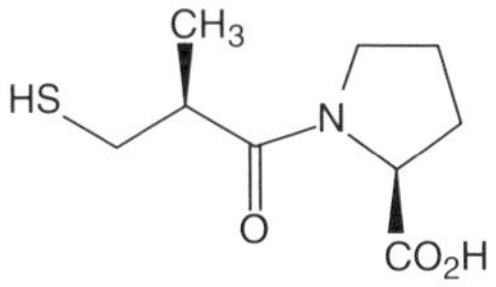

Figure 17   ACE inhibitors.

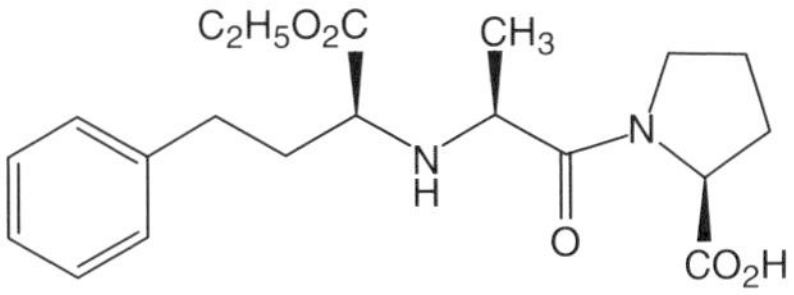

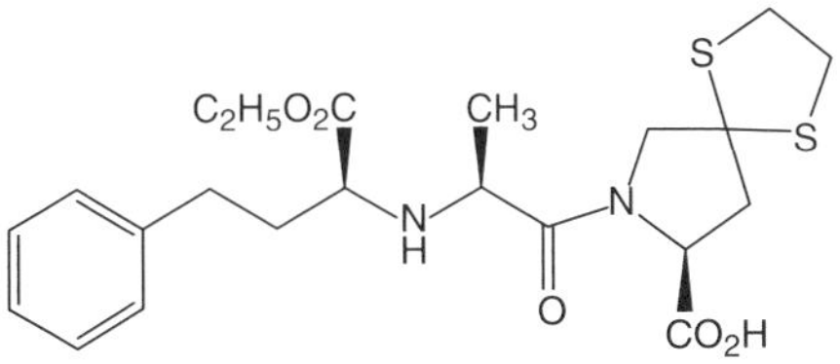

The ACE inhibitors enalapril,[42] spirapril, and quinapril from competing laboratories illustrate the wide diversity of structural modifications that are compatible with this activity. Note, however, that all these compounds retain a similar sequence from the ring nitrogen on. Close to twenty discrete chemical entities have subsequently been assigned nonproprietary names. No fewer than fifteen of those have been approved for use in the United States.

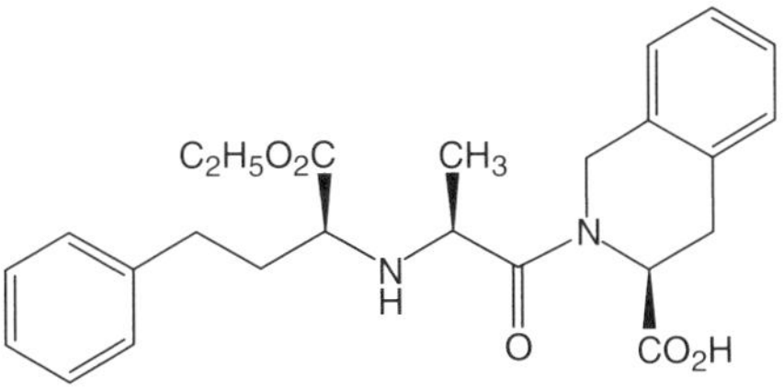

Scientists at Norwich Eaton had also focused on the renin–angiotensin system as a means of developing antihypertensive agents.[46] Instead of seeking ACE inhibitors they set out to find a compound that would antagonize the action of angiotensin II itself by binding with its receptor. This, it was hoped, would inhibit the vasoconstricting action of angiotensin II. By serial substitution of each of the amino acids in the natural substance they developed the antagonist saralasin. The fact that this drug was a peptide precluded widespread use. It was, as expected, not active orally and was quickly metabolized. It does, however, have occasional use by infusion to treat emergency hypertension.

Scientists at the Du Pont–Merck laboratories, in 1990, discovered a small synthetic molecule that acted as an angiotensin antagonist.[47] The compound, later named losartan, had much the same activity as saralasin, although its chemical structure differed significantly from the peptide. One can safely assume that it interacts with the same receptor. The five-membered ring with four nitrogen atoms, a tetrazole, serves the same function as the terminal carboxylic acid in saralasin. The drug offers advantages over the ACE inhibitors as it does not interfere with other targets for converting the enzyme. The discovery was followed relatively soon by angiotensin receptor antagonists from competing firms, eight of which have nonproprietary names and six of which are available in the United States.

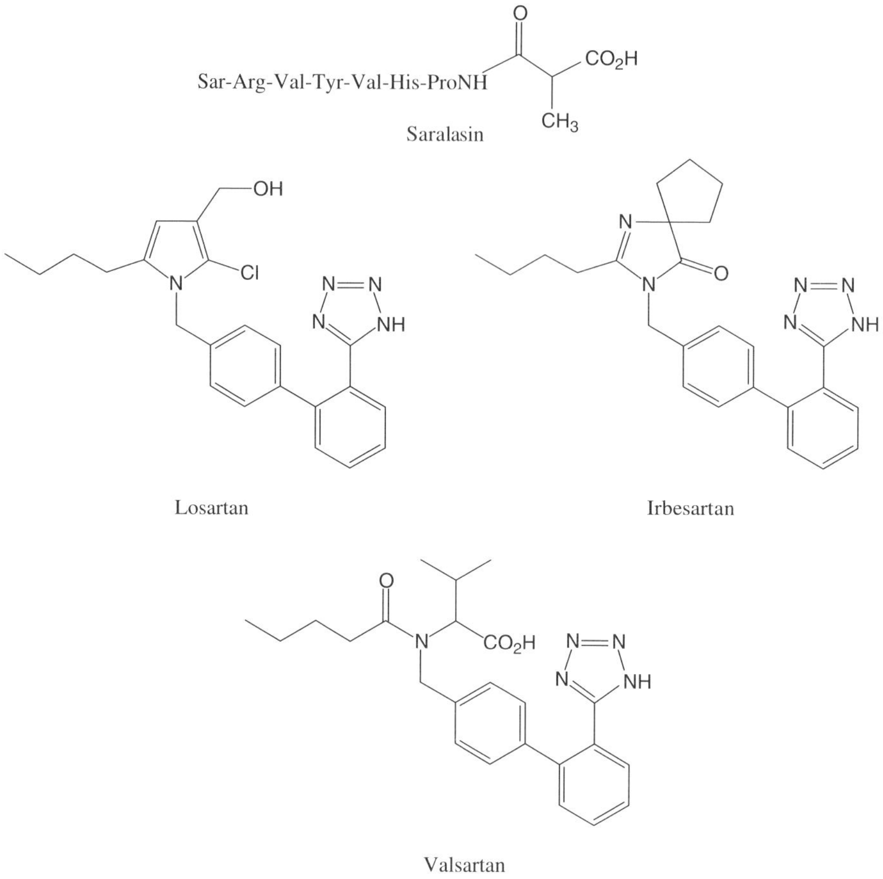

Sar-Arg-Val-Tyr-Val-His-ProNH

Saralasin

Losartan

Irbesartan

Valsartan

Figure 18 Angiotensin antagonists.

The two antagonists depicted in Figure 18, irbesartan[48] and valsartan[49] share the terminal tetrazole ring with losartan.

The central player in the renin–angiotensin system is of course renin itself, the enzyme that cleaves angiotensinogen to begin with (see Fig. 16). Renin belongs to a large class of enzymes known as proteases. These enzymes, as the name indicates, cleave bonds in peptides and proteins to produce smaller peptides. The fermentation product, pepstatin, provided a clue for the design of such small molecule inhibitors. This natural product blocks cleavage of pepsin at a site very similar to that at which renin cleaves angiotensinogen. A very recent publication from the Novartis labs reports on the successful clinical trials of the small-molecule, renin inhibitor aliskiren.[50] The molecule not only blocks the enzyme in vitro, but lowers blood pressure in patients with mild to moderate hypertension (Fig. 19).[51] Another recent compound on the same theme, terlakiren from the Pfizer laboratories, is in an earlier stage of development.[52] It is of note that the very complex structures of

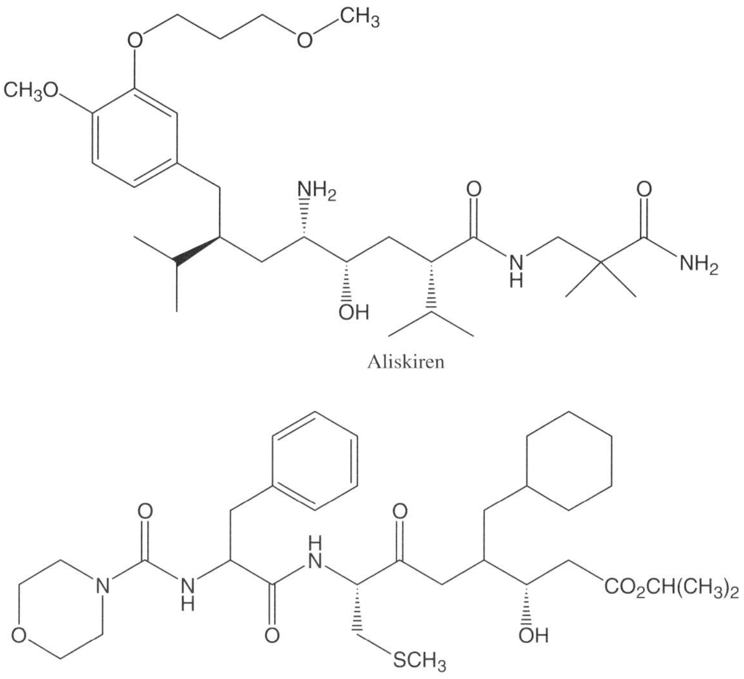

Aliskiren

Terlakiren

Figure 19   Renin inhibitors.

both these renin inhibitors share features reminiscent of small peptides such as the recurrence of amide functions (C=ONH) along the backbone. As noted in Chapter 2, some of the work that led to the development of small-molecule renin directly aided development of a new class of drugs for treating HIV/AIDS patients.

Yet another example of the discovery of a new class of drugs that emerged from the search for new ways in which to treat high blood pressure had its origin in the treatment of asthma, a seemingly quite unrelated disease. It has been known for the better part of a century that the natural product purine base, theophylline, relaxes the constricted bronchioles during asthmatic attacks. The discovery of the role of the phosphodiesterase enzyme (PDE) by Sutherland in the early 1960s helped explain how this drug worked.[53] He and his colleagues had found that a large number of physiological responses were controlled at the cellular level by molecules based on the purine adenosine. The process, simplified, involves the reversible conversion of the esterified form adenosine phosphate (AMP) to cyclized AMP (cAMP). This process, they found is mediated by PDE. Cleavage of the phosphate ester in cAMP by PDE reverses the signal and results in relaxation of involuntary muscle. The development of more delicate assays led to the finding that PDE, like many other messenger molecules, exists as many subtypes. There are recognized at present at least eleven PDEs, each with its own specialized target. It should not come as a surprise that this class of enzymes came under scrutiny as a source of

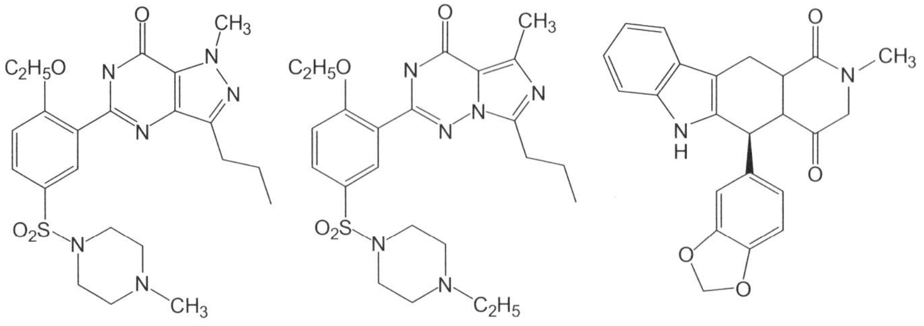

Sildenfanil	Vardenfanil	Tadalafil

Figure 20   Phosphodiesterase 5 inhibitors.

antihypertensive agents (Fig. 20). A PDE selective for arterioles should in theory lower blood pressure. Investigation on the pharmacological properties of these molecules suggested that an antagonist to the subtype PDE 5 should prove particularly useful in treating high blood pressure. A research program at Pfizer led to the discovery of a compound that showed the desired biological action.[54] This was taken to the clinic after it had cleared toxicological tests. Although the compound may have proven its value in treating hypertension, its quite unexpected side effect, reversing erectile dysfunction, soon became the main focus of further studies. The compound, sildenfanil, far better known as Viagra, became a bombshell drug once it reached the market. At least two other competitors, vardenfanil and tadalafil, have been approved for sale in the United States as of this writing. Although the structures of sildenfanil and vardenafil differ only in detail, tadalafil has few features in common with them. This difference may be reflected in its much longer duration of action.[55]

REFERENCES

1. R. B. Barlow and H. R. Ing, *Br. J. Pharmacol. Chemother.*, 153, 586 (1948).
2. H. J. Barber and K. Gaimster, *J. Pharm. Pharmacol.*, 3, 663 (1951).
3. For an overview of early trials see D. G. Beevers, *J. Human Hypertension*, 18, 831 (2004).
4. E. D. Freis, F. A. Finnert, H. W. Schnapper and R. L. Johnson, *Circulation*, 5, 20 (1952).
5. A. Sohn, U.S. patent 2,161,938 (1939).
6. K. Miescher and A. Marxer, U.S. patent 2,503,059 (1950).
7. J. F. Kerwin and G. Ullyott, U.S. patent 2,599,000 (1952).
8. R. M. Cragoe, J. W. Eckholdt and J. G. Wismell, *JAMA*, 202, 870 (1967).
9. G. Brock, *Drugs Today*, 36, 212 (2000).
10. H. Stähle, in J. S. Bindra and D. Lednicer, Eds, *Chronicles of Drug Discovery*, Vol. 1. Wiley, NY, 1982, p. 87.
11. H.-J. Hess, T. H. Cronin and A. Scriabine, *J. Med. Chem.*, 11, 130 (1968).
12. T. H. Althuis and H. J. Hess, *J. Med. Chem.*, 20, 146 (1977).
13. J. W. Constantine, W. K. McShane, A. Scriabine and H. J. Hess, *Postgrad. Med.*, Spec No. 18 (1975).

14. P. B. TIMMERMANS, H. Y. KWA AND P. A. VAN ZWIETEN, *Naunyn Schmiedbergs Arch. Pharmacol.*, 310, 189 (1979).
15. H. ADRIAENSEN AND R. VRYENS, Practioner, 214, 268 (1975).
16. A. S. TURNER, *Br. Med. J.*, 2(6046), 1257 (1976).
17. S. DUTKIEWICZ, A. WITESKA AND K. STEPIEN, *Int. Urol. Nephrol.*, 27, 413 (1995).
18. J. BLACK ET AL., *Br. J. Pharmacol. Chemother.*, 25, 577 (1965).
19. B. N. PRITCHARD, *Br. Med. J.*, 5392, 1227 (1964).
20. For an earlier review see A. M. Barrett, *J. Pharmacol.*, 16, Suppl. 2, 95 (1985).
21. C. I. PHILLIPS, G. HOWITT AND D. J. ROWLANDS, *Br. J. Opthalmol.*, 51, 222 (1967).
22. A. H. NEUFELD, S. P. BARTELS AND J. H. LIU, *Surv. Opthalmol.*, 28, Suppl., 286 (1983).
23. F. C. NOVELLO AND J. M. SPRAGUE, *J. Am. Chem. Soc.*, 79, 2028 (1957).
24. R. W. WILKINS, *New Engl. J. Med.*, 257, 1026 (1957).
25. ANON., *Circulation*, 41, 149 (1970).
26. P. O. ANDERSON AND J. A. KEPLER, *Am. J. Hosp. Pharm.*, 32, 473 (1975).
27. F. SAITO AND G. KIMURA, *Hypertension*, 27, 914 (1996).
28. J. DRUEY AND B. H. RINGIER, *Helv. Chim. Acta*, 39, 195 (1951).
29. M. J. KERT, S. ROSENFELD, J. P. WESTERGRAFT, H. G. CARLETON AND E. HISCOCK, *Angiology*, 5, 318 (1954).
30. A. L TAYLOR ET AL., *New Engl. J. Med.*, 351, 2049 (2004).
31. A. A. RUBIN, F. E. ROTH, M. W. WINBURG, J. G. Topliss, M. H. Sherlock, N. Sperber and J. Black, *Science*, 133, 2067 (1961).
32. G. R. ZINS, D. E. EMMERT AND R. A. WALK, *J. Pharmacol. Exp. Ther.*, 159, 194 (1968).
33. W. C. ANTHONY AND J. J. URSPRUNG, U.S. patent 3,461,461 (1969).
34. G. A. JOHNSON, K. J. BARSUHN AND J. M. McCALL, *Drug Metab. Dispos.*, 11, 507 (1983).
35. W. B. MARTIN, G. R. ZINS AND W. A. FREYBERGER, *Clin. Sci. Mol. Med.*, Suppl. 2, 189 (1975).
36. For a review of topical drugs see G. R. ZINS, *Clin. Dermatol.*, 6, 132 (1988).
37. F. BOSSERT AND W. VATER, *Naturwissenschaften*, 58, 578 (1971).
38. M. THIBONNIER, F. BONNET AND P. CORVOL. *Eur. J. Clin. Pharmacol.*, 17, 161 (1980).
39. J. A. CELLA AND R. C. TWEIT, *J. Org. Chem.*, 83, 4083 (1961).
40. G. DOUGLAS, J. W. HOLLIFIELD AND G. W. LIDDLE, *JAMA*, 227, 518 (1974).
41. For a discussion of this and the development of captopril see M. A. Ondetti, D. W. Cushman and B. RUBIN, in J. S. BINDRA AND D. LEDNICER, Eds, *Chronicles of Drug Discovery*, Vol. 2, Wiley, NY, 1983, p. 1 et seq.
42. A. A. PATCHETT, in D. LEDNICER, Ed., *Chronicles of Drug Discovery*, Vol. 3, ACS Books, Washington, DC, 1993, p. 125 et seq.
43. E. H. GOLD, B. R. NEUSTADT E. M. SMITH, U.S. patent 4,470,972 (1984).
44. S. KLUTCHKO, et al., *J. Med. Chem.*, 29, 153 (1984).
45. P. A. KHAIRALLAH, A. TOTH AND M. BUMPUS, *J. Med. Chem.*, 13, 181 (1970).
46. For a historical overview see D. T. PALS, G. S. DENNING AND R. E. KEENAN, *Kidney Int.*, Suppl. Mar. S7 (1979).
47. J. V. DUNCIA, A. T. CHIU, D. J. CARINI, G. B. GREGORY, A. L. JOHNSON, W. A. PRICE, G. J. WELLS, P. C. WONG, J. C. CALABRESE AND B. M. W. M. TIMMERMANS, *J. Med. Chem.*, 33, 1312 (1990).
48. C. A. BERNHART, B. P. PERRAUT, ET AL., *J. Med. Chem.*, 36, 2182 (1993).
49. P. BUHLMAYER ET AL., *Bioorg. Med. Chem. Lett.*, 4, 20 (1994).
50. P. C. WONG, W. A. PRICE, A. T. CHIU, J. V. DUNCIA, D. J. CARINI, R. R. WEXLER, A. L. JOHNSON AND P. B. TIMMERMANS, *J. Pharmacol. Exp. Ther.*, 252, 726 (1990).
51. A. H. GRADMAN, R. E. SCHMIEDER, R. L. LINS, J. NUSSBERGER, Y. CHIANG AND M. P. BEDIGIAN, *Circulation*, 111, 101 (2005).
52. M. L. MANGIAPANE, R. T. WEBSTER, W. F. HOLT, S. S. ELLERY, K. A. SIMPSON, T. M. SCHELHORN, A. H. SMITH, I. M. PURCELL AND W. R. MURPHY, *Drug Develop. Res.*, 34, 361 (2004).
52. See, for example, V. STEFANOVICH, M. VON POLNITZ AND M. REISER, *Arzneim.*, 11, 1747 (1974).
53. N. K. TERRETT ET AL., *Bioorg. Med. Chem. Lett.*, 6, 1819 (1996).

Chapter 4

Lipid Lowering Drugs

Circulatory problems resulting from abnormal deposition of cholesterol and triglycerides, collectively termed lipids, comprise a second major risk factor in cardiovascular disease. The preceding chapter focused mainly on hypertension as the other major risk factor. The discussion centered on drugs that lowered blood pressure via the set of nervous and muscular controls of the vasculature that adjust blood pressure. Lipid deposition on the vessel walls plays yet another part in compromising blood circulation. This disease, usually called arteriosclerosis, has been described as the leading cause of illness and death in the United States. The terms arteriosclerosis and atherosclerosis are often used interchangeably; the latter, which is derived from a description of the deposits within the blood vessels (athero: from the Greek porridge) is actually technically a subclass of arteriosclerosis, this term denoting hardening (sclerosis) of the arteries. Atherosclerosis has been called "the most important underlying disease in the industrial world leading to: coronary heart disease, cerebrovascular disease and peripheral vascular disease."[1]

The problem arises from the fact that triglycerides and cholesterol are essential to various life processes. The metabolism of the fatty acids that make up the triglycerides, more familiarly known as fats, are used by most living organisms as a source of energy. Diet is the principal source of these substances. The other constituent of lipids, cholesterol, plays a multitude of roles in maintaining life, serving not only as a starting material for the synthesis of the host of steroid hormones, but also forming part of various structural membranes. Triglycerides and cholesterol must often be conveyed from one site in the organism to another via the circulatory system. The fact that these fatty substances are virtually insoluble in blood, which comprises mainly water, calls into action a set of special carrier molecules, the so-called lipoproteins. When freshly ingested, or synthesized, lipids first combine with those carriers, forming very low density lipoproteins (VLDL). As these age, these are converted to low density lipoproteins (LDL) by shedding some of the triglyceride load; these LDL are also the main carriers of cholesterol. A third category of

New Drug Discovery and Development by Daniel Lednicer
Copyright © 2007 John Wiley & Sons, Inc.

transport proteins comprises the high density lipoproteins (HDL), one of whose main functions involves carrying cholesterol from various sites back to the liver where it is metabolized or stored. These are familiarly classed in the popular press as "bad cholesterol" (VLDL and LDL) and "good cholesterol" (HDL).

Atherosclerosis is a slow progressive process that may actually start early in life. One of the two contending theories proposes that the series of events begins with an injury to the lining of the blood vessels. This causes some of the fat-laden lipoproteins to accumulate just below the vessel wall. The ensuing deposit attracts further lipid-loaded proteins. The site eventually narrows the blood vessels and may form local calcified plaques. The lipid hypothesis on the other hand assigns the origin of the process to elevated blood lipid levels.[2] The net result, whatever the origin, leads to decreased flexibility of the arterial wall, familiarly known as "hardening of the arteries," decreased vessel diameter, and formation of calcified plaques. The first two sequences result in the diminished blood flow that may be manifested in a variety of symptoms depending on the location of the sclerosis; this is often felt as pain resulting from lack of oxygen in blood-deprived muscle. This shows up as angina from narrowed coronary arteries. Pain on walking in peripheral vascular disease is due to narrowing of the arteries in the legs. The plaques that form in blood vessels as a result of atherosclerosis are more immediately life threatening. These often travel to crucial sites when they break off; once there, the plaques can totally shut off the supply of blood to a major organ. This deprives the tissue of the oxygen required for survival. Many heart attacks and most strokes are direct results of such events.

The involvement of cholesterol in coronary artery disease was discovered surprisingly early. This compound, whose structure was not known at the time, was identified in plaques in blood vessels by a pathologist in 1845, over a century and a half ago. Experiments that produced atherosclerosis in rabbits by feeding the animals cholesterol were described in 1913. A by now classic epidemiological study provided evidence that the connection between heart disease and cholesterol levels applied to humans as well. The Framingham Heart Study was begun in the small Massachusetts town of the same name in 1948. The project initially recruited a group of over 5000 individuals with the intent of studying their cardiovascular health over a lifetime. The study is still ongoing and has been extended to additional groups and third-generation descendants. As early as 1960, data from the first group had confirmed that elevated cholesterol levels increased the risk of heart disease. It was assumed by many pharmaceutical researchers that lowering of the high levels would decrease that risk. Projects were thus launched in a number of laboratories to find drugs that would lower serum cholesterol.

Nicotinic acid is one of the forms of niacin, the vitamin in the B complex associated with prevention of pellagra. A report that higher than physiological doses of this structurally simple compound lowered elevated cholesterol levels was published[2,4] in 1965, just a few years after the Framingham finding. This drug, which is still used today, does however cause some annoying side effects such as hot flushes. It has recently been found that the drug not only lowers LDL, but also causes a corresponding increase in HDL. The latter effect is believed to

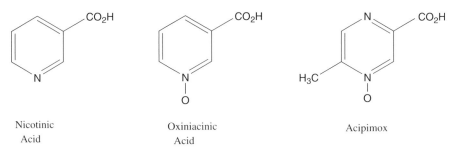

Nicotinic
Acid

Oxiniacinic
Acid

Acipimox

Figure 1   Nicotinic acid and its derivatives.

further decrease the risk from atherosclerosis. The various synthesis programs that used nicotinic acid as their lead have led to very few other derivatives (Fig. 1). This includes the simple oxidation products oxiniacinic acid and acipimox.[3] This last is based on a pyrazine ring system that includes an additional nitrogen. That drug is licensed for sale in the UK.

Cholesterol is obtained both by ingestion of food and by synthesis from acetate in the body. Endogenous synthesis normally accounts for the larger proportion of drug present in the body. A more detailed discussion of that process will come up in connection with the "statins." The level of blood cholesterol is normally determined by an active equilibrium that involves compound from synthesis and diet at one end and metabolic disposition at the other. This last process occurs in the liver. This organ carries out an oxidation process in which cholesterol is converted to water-soluble bile acids such as, among other metabolites, cholic acid (Fig. 2). These are then conveyed to the bile duct from whence they flow into the intestine to be eventually excreted in feces.

The level of metabolic activity, called the basal level, is dictated by the concentration of the hormone thyroxin (Fig. 3). The blood concentration of this compound also controls the rate at which the liver oxidizes various circulating compounds to a form suitable for excretion. Increased thyroxin levels thus lead to a higher rate of disposition of cholesterol; this then results in lower serum levels

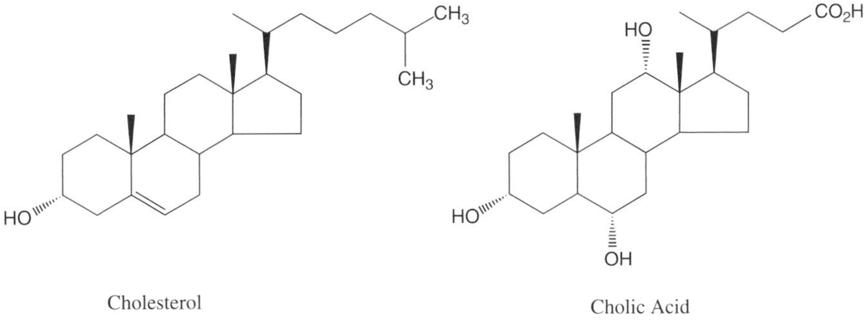

Cholesterol

Cholic Acid

Figure 2   Oxidation of cholesterol to cholic acid.

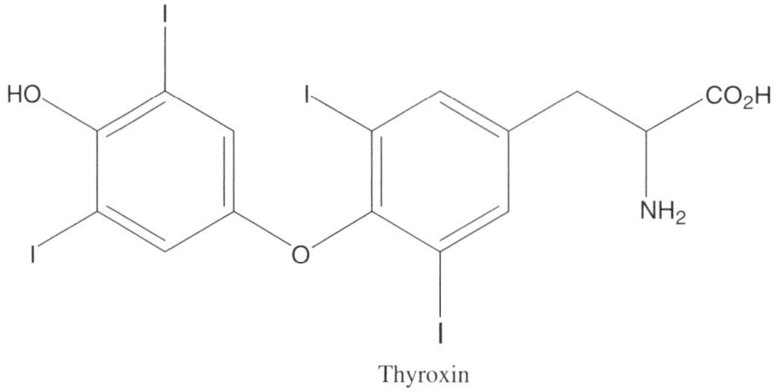

Thyroxin

Figure 3   Thyroxin.

of that compound. Because of its central role in a host of metabolic processes, thyroxin cannot, however, be used as a drug to regulate the isolated endpoint of cholesterol levels. The compound, however, exists in two mirror-image forms as a consequence of the presence of four different substituents on the carbon bearing the amino group. Virtually all the metabolism regulating activity of thyroxin comes from a single mirror form, the L isomer. The other isomer, D-thyroxin, although largely devoid of most other regulatory functions, does retain the lipid-lowering activity. This compound is approved for that indication, although directions for its use are studded with a host of cautionary remarks on possible other metabolic effects.

Another way to accomplish the same effect as D-thyroxin, that is, to increase the metabolism of cholesterol, involves enhancing the rate at which the oxidation products are excreted. This strategy relies on the existence of a feedback mechanism for maintenance of levels of bile acids. Removal of bile acids should, at least in principle, lead to increased rate of oxidation of cholesterol to bring the level of those acids back up to some preset concentration. This should in theory in the end lead to a drop in levels of cholesterol. It was in fact found in 1960 that feeding ion exchange resins to laboratory animals lowered serum cholesterol.[5] The basic polymers that were administered orally proceeded into the intestine unchanged as they were very insoluble and thus not absorbed into the circulation. There, the resins bound bile acids secreted into the gut from the bile duct by an ordinary acid–base interaction. The bound cholesterol oxidation products then proceeded down the intestine and were finally excreted. This confirmation that serum cholesterol levels could be controlled by sequestering bile acids in the gut led within a few years to a practical product. The active compound of this drug comprises polystyrene cross-linked with divinylbenzene, which in turn is linked covalently with a strongly basic trimethylammonium-methyl group. Figure 4 depicts a typical small fragment of the very large structure as though it were flat; the actual structure is an intricate three-dimensional network. The original batches used in animal experiments may well have consisted of one of the commercial Dowex® ion exchange resins that have much the same composition. Clinical trials revealed that this compound, by now named cholestyramine, was an effective cholesterol-lowering agent in humans.[6] Some years ago when the drug was still only available

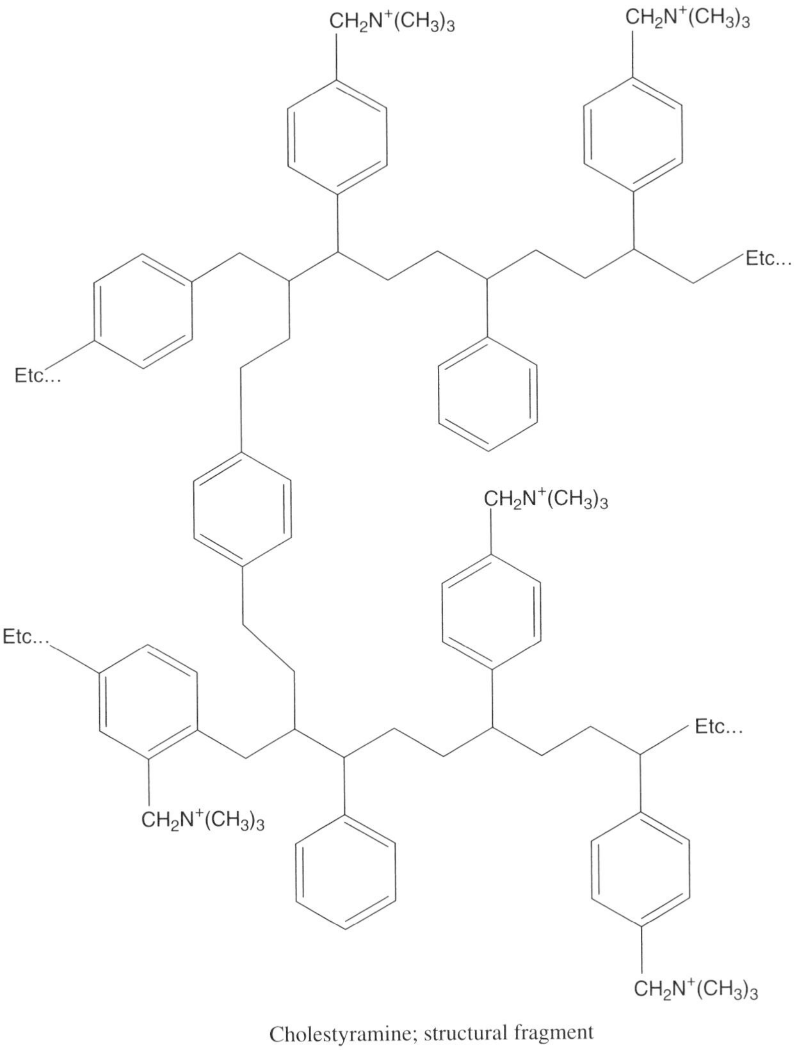

Cholestyramine; structural fragment

Figure 4   Cholestyramine structural fragment.

from a single company as Questran®, the active drug substance came from the same production line as that of the Dowex® resins. When a new batch of drug was required, the manufacturer, Dow, shut down production and subjected the plant to a thorough cleaning. That done, they put through a batch of pharmaceutical-grade resin. The very large dose required, 10–20 g per day, and the unpleasant mouth feel was a significant drawback to this quite effective drug. The more recently introduced tablet form overcomes some of the dosing problem.

The introduction of cholestyramine led to a limited amount of work at other firms to develop their own cholesterol-lowering resins. Arguably in the hope of finding a resin that was effective at lower doses, chemists at Upjohn concentrated on polymers based on polyamines, that is, long-chain compounds in which every third atom is nitrogen. The products, whose structures in many ways resembled those of epoxy resins, should in theory be able to complex more bile acids per unit dose because of the higher proportion of basic nitrogen atoms per gram than present in cholestyramine. Scientists at Upjohn indeed found that a polymer obtained from a commercial polyamine product called tetraethylene pentamine and the epoxide epichlorohydrin effectively lowered serum cholesterol in experimental animals.[7] Data from clinical trials on this agent, by now named colestipol, showed that this activity persisted in humans as well.[8] As the product was submitted for regulatory approval it became apparent that the name for one of the starting materials, "tetraethylene pentamine," was actually a label writer's fancy. The material in fact comprised a very complex crude mixture of polyamines. Replacement of that starting material by triethylene tetramine, a substance available as a pure chemical, gave a resin[9] that had exactly the same biological properties in animals as well as in humans. Figure 5 depicts the starting materials as well as an attempt to depict a small unit of the very complex three-dimensional tangle that comprises colestipol. The hope for higher potency was not realized; the prescribed dose of colestipol too was measured in terms of tens of grams per day. This drug too is now available in tablet form.

Research on atherosclerosis in the late 1950s had produced some evidence that the steroid androsterone, a distant metabolite of testosterone, lowered elevated serum lipids. This compound was, however, difficult to administer because of its very poor solubility in any media. An apocryphal account has it that scientists at

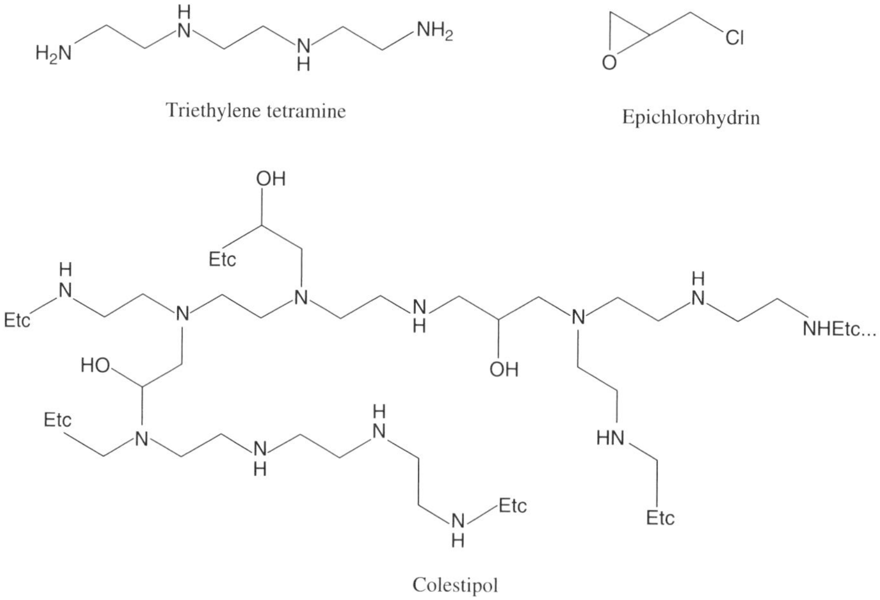

Triethylene tetramine   Epichlorohydrin

Colestipol

Figure 5   Colestipol monomers and structural fragment.

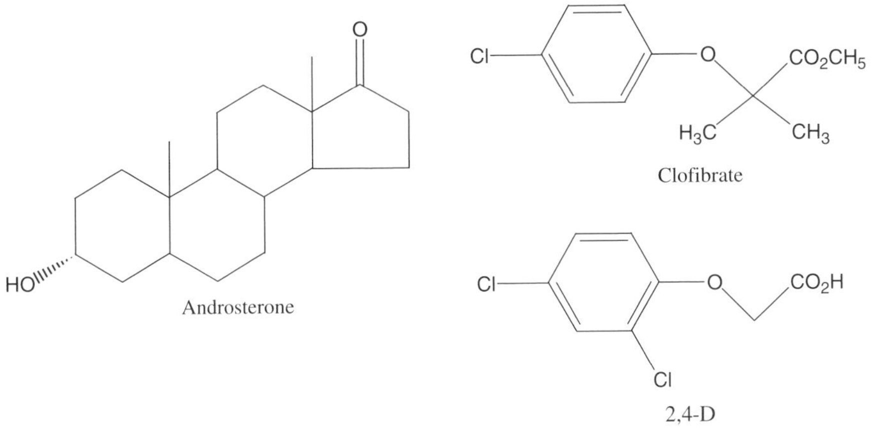

Figure 6 Atromid and its antecedents.

ICI searched their collection of oily synthetic compounds for a suitable solvent. The chemical structure of the agent chosen, clofibrate, bears a distinct relation to synthetic herbicides such as 2,4-dichloropenoxyacetic acid, more familiarly known as 2,4-D. The combination of the steroid and clofibrate, given the name Atromid, proved to be a reasonably effective lipid-lowering drug (Fig. 6). After it was marketed in the early 1950s, a number of investigators tested both the combination and what was originally thought to be the solvent separately in both animal models of atherosclerosis and in humans. These trials fairly unequivocally showed that the lipid-lowering activity was due entirely to the vehicle, which was then named clofibrate.[10] This finding led to the introduction of this last as a drug in its own right. The name Atromid S assigned to that product indicates its provenance, S being the pharmacist symbol for without as in the Latin *sine*. Recent research supports the hypothesis that clofibrate is first hydrolysed to the corresponding acid, clofibric acid, in serum. This acid lower lipids by enhancing triglyceride hydrolysis. This in turn reduces the number of lipid-rich LDL and VLDL particles in blood.

The drugs that preceded Atromid S, nicotinic acid, D-thyroxin and the bile acid sequestering resins, had serious shortcomings; they either had side effects that could be considered an extension of their mode of action or else had to be taken in very large doses. The relatively good tolerance of this new drug led to ready acceptance in the market. This not unexpectedly led to competitive firms coming up with their own versions. As a result, the Merck Index lists no fewer than a dozen agents in this class, with more or less closely related chemical structures, that have been assigned nonproprietary names (Fig. 7). These drugs, collectively known as "fibrates," all lower lipids by the same mechanism as clofibrate itself.[11] Currently approved drugs in this class include bezafibrate,[12] ciprofibrate,[13] fenofibrate,[14] and gemfibrozil.[15] These are often prescribed for use along with a lipid-lowering agents selected from a different mechanistic category.

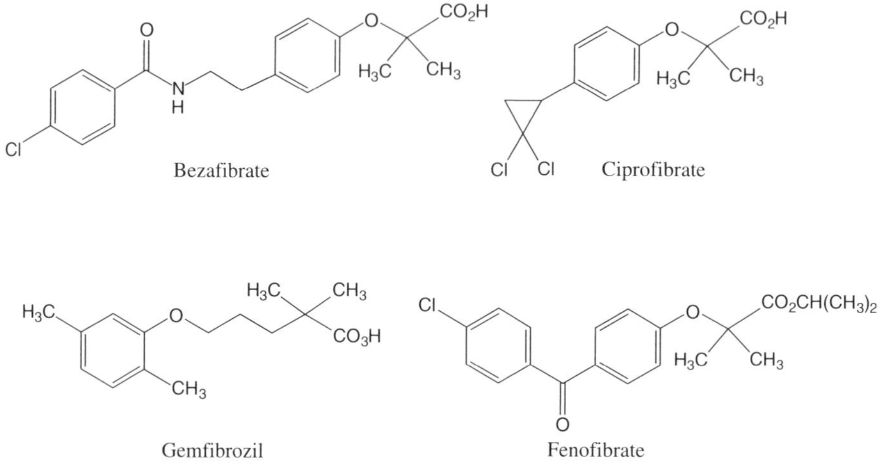

Figure 7   Selected clinical fibrates.

Cholesterol, whether found in animals or in plants, is the end product of a lengthy chain of synthesis reactions carried out by cells within the organism. The basic structural unit in that scheme is the equivalent of a simple, unsaturated five-carbon unit that comprises a straight chain branched at one end. In plants this unit, isoprene, can then go on to form a host of natural products called terpenes, which are made up from two to as many as six isoprene fragments linked in various ways. The process is quite similar in animals, proceeding initially to the terpene lanosterol made up of six isoprenes. This compound has the four fused-ring steroid ring system; several additional steps finally lead to cholesterol (Fig. 8). It should be possible to decrease cholesterol levels by inhibiting endogenous synthesis by interrupting one of the steps in the long sequence. One of the very first steps in the sequence involves formation of the isoprene equivalent, isopentenyl pyrophosphate. The key reaction in that chain actually controls the rate at which the process proceeds; interruption at this stage will directly halt the process that leads to cholesterol. That early step comprises reduction of the activated carboxylic acid group (COSCoA) in the glutaric acid derivative hydroxymethylgluraryl CoA to an alcohol (CH_2OH); the product, mevalonic acid, then goes on to the isoprene equivalent, isopentenyl pyrophosphate, in several more steps. That pivotal reaction is catalysed by an enzyme that goes by the daunting name hydroxymethylglutaryl CoA reductase. The acronym for that enzyme HMG-CoA would give the name to an important class of cholesterol lowering agents, the HMG-CoA inhibitors.

By the 1970s, the search for new antibiotics from fungi was yielding diminishing returns. Advances in other fields had led to the identification of many enzyme systems that were associated with various diseases. The role of HMG-CoA in the synthesis of cholesterol, for example, as outlined above, had by then been well established; an inhibitor of this substance was an attractive screening target. The fact that an enzyme preparation was available offered the possibility of testing

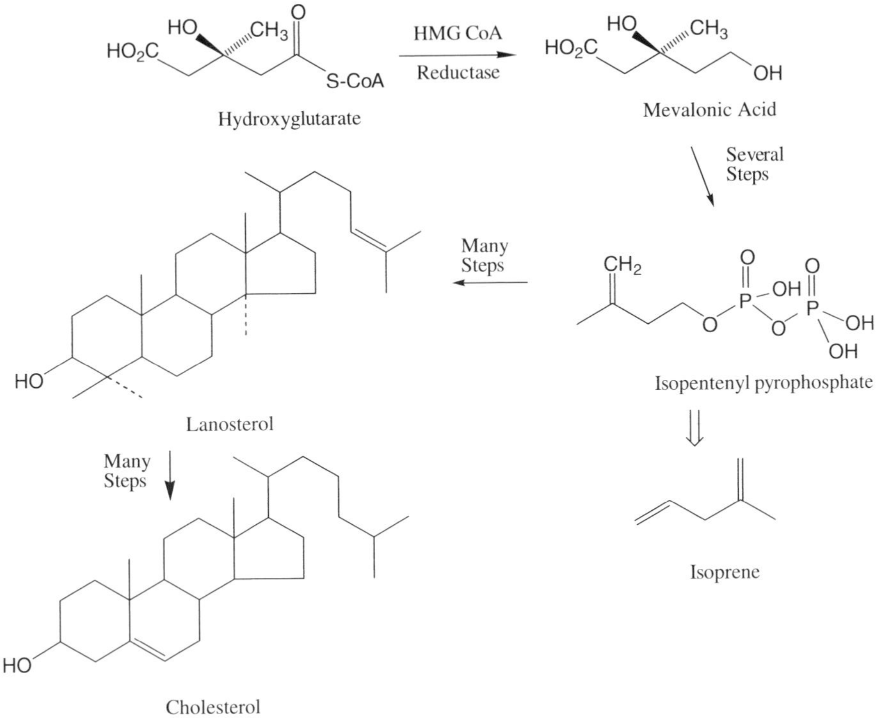

Figure 8   From hydroxyglutarate to cholesterol: the role of HMG-CoA.

agents as inhibitors in vitro at very low concentrations, whether as pure compounds or as parts of very crude mixtures. Endo and Kuroda at the Sankyo laboratories set up a program to test fermentation mixtures against that enzyme, reasoning that some organisms might produce an inhibitor in defense against other microbes.[16] In 1976, after a long search, they reported the identification of an HMG-CoA inhibitor from a culture of *Penicillium citrinum*.[17] They named the compound mevastatin to indicate that it inhibited (*statin*) the synthesis of mevalonic acid (*meva*). This paper was followed not much later by a report by scientists at Merck of the isolation and structure determination of a fermentation product that had very similar activity on inhibiting cholesterol synthesis. They named this compound compactin after the producing organism *Penicillium brevicompactum*.[18] This product was eventually shown to be identical to mevastatin. The structure of this first statin interestingly incorporates a mevalonate-like structural fragment as a cyclic internal ester attached to the right-hand ring by a carbon–carbon bond. This piece, called a lactone, looks more like mevalonic acid after it opens to the parent hydroxyl-acid by hydrolysis. That structural fragment was to prove an essential structural feature for activity for all the subsequent HMG-CoA inhibitors, a class that was to be known collectively as the "statins." This portion of the molecule causes the statins to occupy a site on the enzyme normally occupied by glutarate.

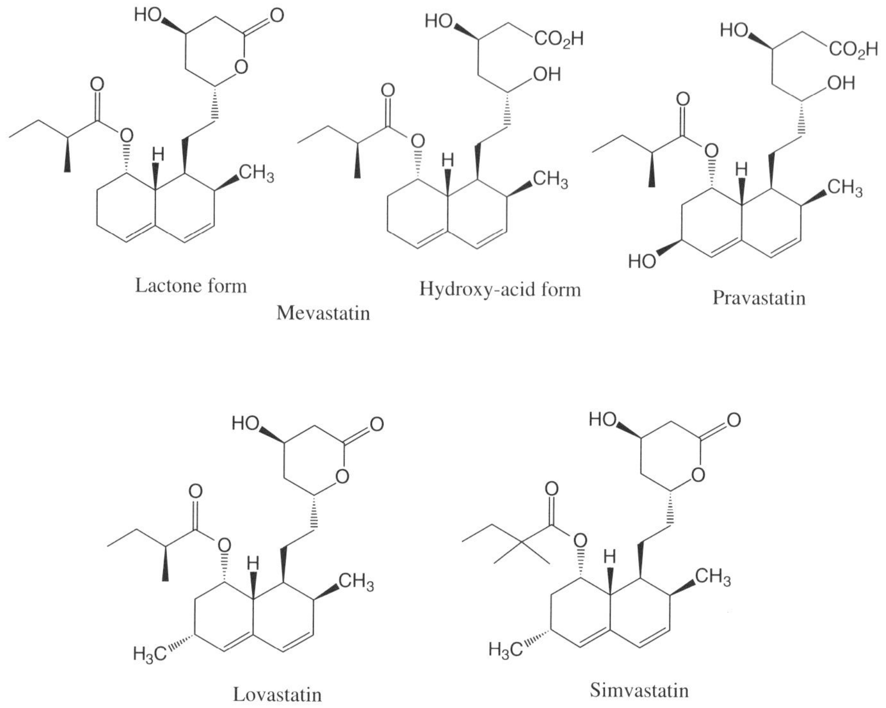

Lactone form

Hydroxy-acid form

Pravastatin

Mevastatin

Lovastatin

Simvastatin

Figure 9   Natural product statins.

This in essence deprives the enzyme of fresh glutarate starting material for conversion to mevalonate.[19]

Lovastatin was the first of the statins to be introduced into clinical practice (Fig. 9). This fermentation product differs from the lead compound, mevastatin, by the presence of an additional methyl group in one of the fused rings.[20] A second fermentation of mevastatin with a different organism introduces an additional hydroxyl group into the same ring.[21] The resulting compound, pravastatin, is one of the more prominent of the currently available statins. Partial synthesis gained a very prominent role in the history of antibiotics leading for example to the modified cephalosporins. A similar strategy led to the new statin simvastatin. This compound is prepared by replacing the ester side chain present in the other statins with one that has an additional methyl group.[22]

These fermentation-derived statins share a mevalonic acid-like fragment attached to an all-carbon, six-membered fused-ring portion, called a decalin. This portion of the molecule is possibly involved in transport of the drug to the site of action; it probably also plays a role in fitting the molecule on the HMG-CoA binding site. Research in several competing firms concentrated on compounds that were prepared by synthesis in the laboratories of organic chemists instead of fermentation-derived products. This work revealed that fragments that differed widely from the decalin would fulfill the same functions as the decalin. The results from

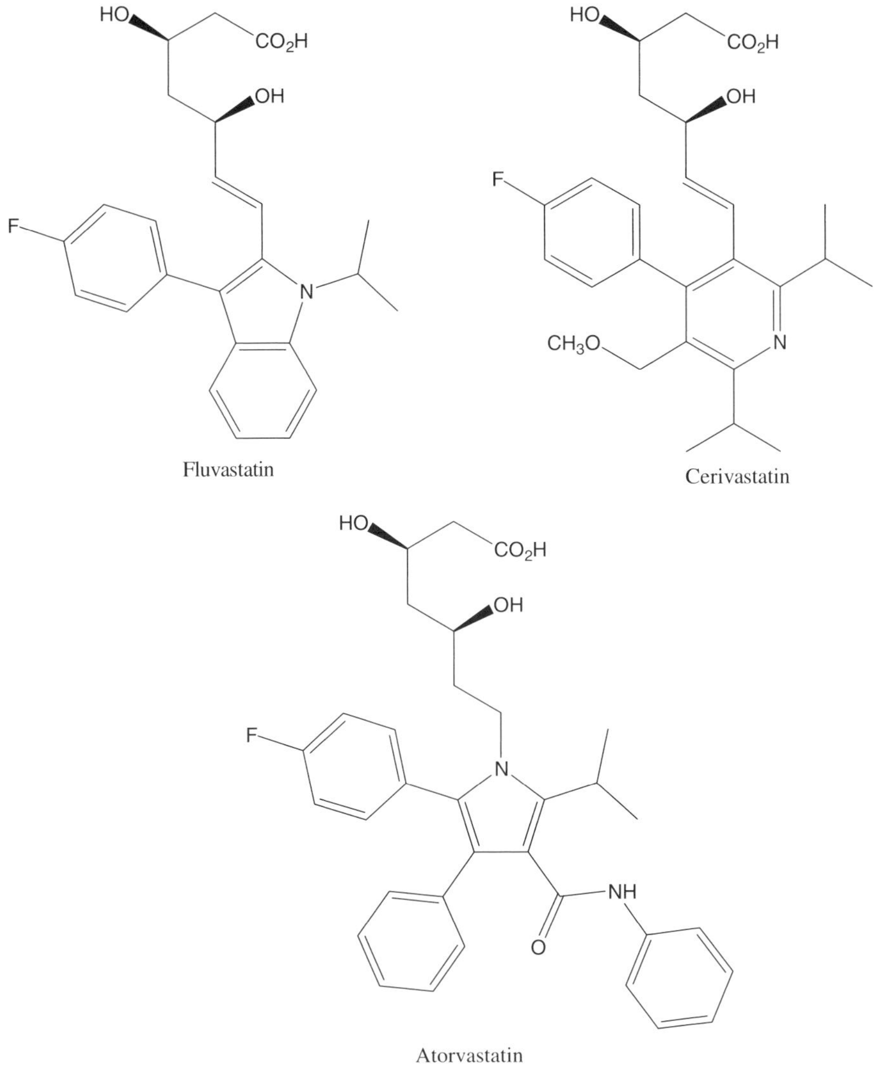

Fluvastatin

Cerivastatin

Atorvastatin

Figure 10 Synthetic statins.

that work led to a series of products whose preparation required a somewhat lengthy and intricate series of chemical transformations. The mevalonate recognition group is, for example, attached to a nitrogen-containing indole ring in the drug fluvastatin (Fig. 10).[23] The quite complex compound atorvastatin[24] carries that group directly on the nitrogen atom of a pyrrole ring. The mevalonate-like group in the drug cerivastatin[25] is attached to a six-membered nitrogen-containing pyridine ring. The widely differing chemical structures of the decalin surrogates present in these drugs may mean that the compounds are distributed in different ways to parts of the body. They thus tend to show somewhat different pharmacological profiles and patterns of side effects from the fermentation statins and from each other. A rare side effect observed among patients who were taking cerivastatin, as an example, led Bayer, in 2001, to voluntarily withdraw the drug, marketed as Baycol®, from the U.S. market.

The widespread use of these drugs reflects their efficacy in lowering elevated cholesterol levels while at the same time increasing HDL levels. There is additional evidence that they have beneficial effects in treating coronary artery disease that may be unrelated to lowering serum cholesterol levels.[26] Unconfirmed accounts report beneficial effects on inflammatory conditions and even Alzheimer's disease.

The amount of cholesterol produced by body cells makes up only one portion of that compound that shows up in the HDL and LDL lipoproteins. The "Nutrition Facts" box label on each container of food is a reminder that diet contributes a significant quantity of the cholesterol that in the end shows up in serum. Advice on cardiovascular health more often than not includes admonitions to avoid cholesterol-rich foods. As this is not always possible, some attention has been devoted over the years to finding means for limiting the absorption of ingested cholesterol. The plant sterol beta-sitosterol, which differs in structure from cholesterol only by some slight changes in the side chain, had at one been tried for that indication. It was reasoned that the drug might competitively block absorption of cholesterol. Absorbed sitosterol would not pose a problem because it was believed to be biologically inert. This compound is still advertised for that purpose by various nutritional supplement sources, although the effect on cholesterol levels ranges from nil to at best very modest.

Cholesterol is not absorbed from the intestine as such, but needs first to be esterified. This process requires a special enzyme, ACAT (acyl-CoA : cholesterol acyltransferase). Scientists at Schering–Plough launched a program in the early 1990s aimed at identifying compounds that would block that enzyme and thus inhibit absorption of cholesterol. They prepared a series of analogs guided by an in-vitro assay designed to find inhibitors of the enzyme. The most active of these was a compound based on a four-membered nitrogen ring, a so-called azetidone.[27] This agent, given the code number SCH 48461, effectively lowered cholesterol absorption in an animal model. Follow-up experiments, however, showed that the compound was quickly and extensively metabolized as soon as it was administered. The fact that the metabolite was more active than the drug itself, a circumstance reminiscent of minoxidil (see Chapter 3), led them to prepare additional analogs. One of the methoxyl groups (CH_3O) in the metabolite had been cleaved to a free phenolic hydroxyl (OH). They thus prepared a second series of analogs, all of which incorporated that feature. The compound selected for clinical trial, ezetimibe, is up to four hundred times more potent than SCH 48461 in some assays (Fig. 11).[28] The drug proved to be an effective inhibitor of cholesterol absorption in humans as well. It was approved for use in humans in late 2002. Although the initial lead came by way of a search for ACAT inhibitors, the actual mechanism by which this compound inhibits absorption of cholesterol is not yet fully understood. A fixed combination tablet of ezetimibe with simvastatin was recently approved and

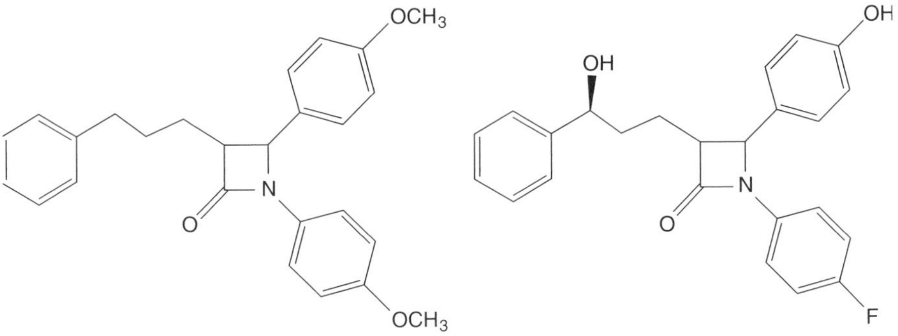

SCH 48461   Ezetimibe

Figure 11   Ezetimibe and its predecessor.

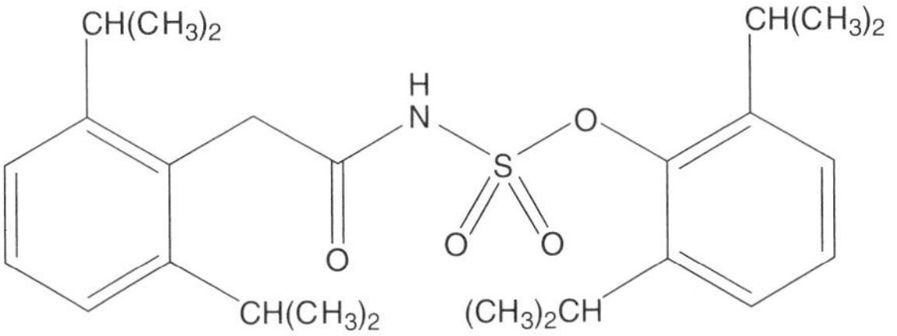

Figure 12   Avasimibe.

launched on the market in 2005. This drug should lower serum cholesterol levels by both decreasing synthesis and inhibiting absorption of that compound from the diet.

The search for agents that would lower serum cholesterol by inhibiting the ACAT enzyme led chemists at Warner–Lambert to a series with markedly different chemical structures from the foregoing. This group's research led to the ACAT inhibitor avasimibe,[29] a compound that incorporates an unusual functional group

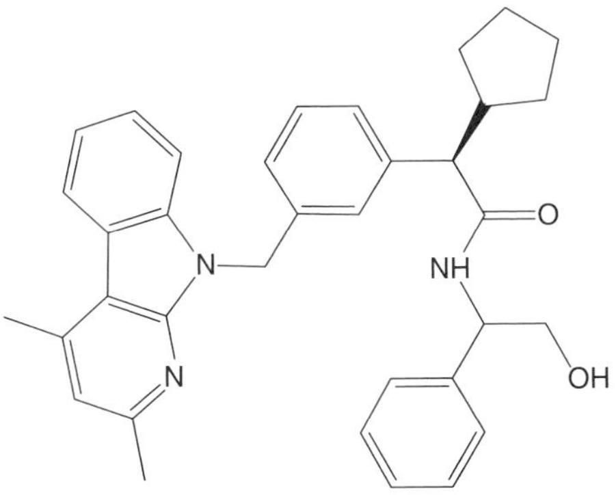

Figure 13   Implitapide.

describe as an acyl sulfamate (Fig. 12). The standard animal model in much of the research described thus far involves measuring the effect of drugs in animals that are fed diets laced with high levels of cholesterol. Avasimibe notably lowered serum levels of cholesterol even in animals that were fed nonenriched normal diets. The compound may have a direct effect on atherosclerosis in addition to its lipid-lowering activity, and has also shown lipid-lowering activity in humans.[30]

A protein with the acronym MTP (microsomal transfer protein) transfers lipids between various reservoirs. The fact that the substance plays a key function in the assembly of lipoproteins makes it yet another target for the design of agents that can potentially control elevated lipid levels. The structurally very complex compound implitapide (Fig. 13), from the Bayer labs, has been shown to inhibit MTP function in laboratory animals.[31,32]

REFERENCES

1. W. KANNEL AND P. W. F. WILSON, in A. GAW, C. PACKARD AND J. SHEPHERD, Eds, *Statins: The HMG COA Reductase Inhibitors in Perspective*, 2nd ed., Taylor & Francis, London, 2004, pp. 19–31.
2. ANON., *Merck Manual of Diagonsis and Therapy*, Section 16, Chapter 201.
3. H. E. J. THOMAS, W. B. KANNEL, T. R. DAWBER AND P. M. MCNAMARA, *N. Engl. J. Med.*, 274, 701 (1966).
4. L. A. CARLSON AND L. ORO, *Acta Med. Scan.*, 172, 641 (1962).
5. D. M. TENNENT, H. SIEGEL, M. E. ZANETTI, G. W. KURON, W. R. OTT AND F. J. WOLF, *J. Lipid Res.*, 469 (1960).
6. S. A. HASHIM AND T. B. VAN ITALLIE, *J. Am. Med. Assoc.*, 192, 289 (1965).
7. T. M. PARKINSON, K. GUNDERSON AND N. A. NELSON, *Atherosclerosis*, 11, 531 (1970).
8. E. R. NYE, D. JACKSON AND J. D. HUNTER, *N. Z. Med. J.*, 76, 12 (1972).
9. D. LEDNICER AND C. Y. PEERY, U.S. patent 3,803,237 (1974).
10. P. CARSON, L. MCDONALD, S. PICKARD, T. PILKINGTON, B. DAVIES AND F. LOVE, *Br. Heart J.*, 28, 400 (1966).
11. J. AUWERX, K. SCHOONJANS AND B. STAELS, *J. Atheroscler. Thromb.*, 3, 81 (1996).
12. E.-C. WITTE, K. STACH, M. THIEL, H. SCHMIDT AND H. STORK, U.S. patent 3,781,328 (1973).
13. D. K. PHILLIPS, U.S. Patent, 3,948,973 (1974).
14. P. L. CREGER, G. W. MOERSCH AND W. A. NEUKLIN, *Proc. Roy. Soc. Med.*, 69, 3 (1976).
15. A. MIEVILLE, U.S. patent 4,058,552 (1977).
16. For a review see A. ENDO, in A. Gaw, C. PACKARD AND J. SHEPHERD, Eds, *Statins: The HMG COA Reductase Inhibitors in Perspective*, 2nd edn, Taylor & Francis, London, 2004, p. 32 et seq.
17. A. ENDO, *J. Antibiotics (Japan)*, 29, 1346 (1976).
18. A. G. BROWN, T. C. SMALE, T. J. KING, R. HASENKAMP AND R. H. THOMPSON, *J. Chem. Soc.*, [Perkin 1], 1165 (1976).
19. E. S. ISTVAN AND J. DEISENHOFER, *Science*, 292, 1160 (2001).
20. A. ENDO, *J. Antibiotics (Japan)*, 32, 852 (1979).
21. N. SERIZAWA, S. SERIZAWA, K. NAKAGAWA, K. FURUYA, T. OKAZAKI AND A. TERAHARA, *J. Antibiotics (Japan)*, 36, (1983).
22. W. F. HOFFMAN, R. L. SMITH AND A. K. WILLARD, *J. Med. Chem.*, 29, 849 (1986).
23. J. PROUS AND J. CASTANER, *Drugs Fut.*, 16, 104 (1991).
24. K. L. BAUMANN, D. E. BUTLER, C. F. DEERING, K. E. MENNEN, A. MILLAR, T. N. NANNIN, C. W. PALMER AND B. D. ROTH, *Tetrahedron*, 33, 2283 (1992).
25. R. ANGERBAUER, W. BISCHOFF, W. STEINKE AND W. RITTER, *Drugs Fut.*, 19, 537 (1994).
26. W. PALINSK, *Arteriosclerosis, Thrombosis and Vasc. Biol.*, 21, 3 (2001).

27. D. A. BURNETT, M. A. CAPLEN, H. R. DAVIS, R. E. BURRIER AND J. W. CLADER, *J. Med. Chem.*, 37, 1733 (1994).
28. S. B. ROSENBLUM, T. HUYN, A. AFONSO, H. R. DAVIS, N. YUMIBE, N. CLADER AND D. A. BURNETT, *J. Med. Chem.*, 41, 973 (1998).
29. H. T. LEE, D. R. SLOSKOVIC, J. A. PICARD, B. D. ROTH, W. WIERENGA, J. L. HICKS, R. F. BOUSLEY, K. L. HAMELEHLE, R. HOMAN, C. SPEYER, R. L. STANSFIELD AND B. R. KRAUSE, *J. Med. Chem.*, 39, 5031 (1996).
30. G. LLAVERIAS, J. C. LAGUNA AND M. ALEGRET, *Cardiovasc. Drug Rev.*, 21, 33 (2003).
31. L. A. SORBERA, L. MARTIN, J. SILVESTRE AND J. CASTANER, *Drugs Fut.*, 25, 1121 (2000).
32. M. SHIOMI AND T. ITO, *Eur. J. Pharmacol.*, 43, 127 (2001).

Chapter 5

Centrally Acting Analgesics

Webster's Dictionary devotes a full column of exceedingly fine print to definitions of the word *pain*. In spite of this, humans find it only too easy to recognize that discomfiting sensation. Pain of course plays an important role in our relation with the environment. It can be considered the first and most direct response to a potentially injurious circumstance. Almost every explanation of the term invokes the reaction to touching a finger to a hot stove. The painful sensation that results from contact leads to quick withdrawal of the finger and thus avoidance of a burn. Pain-causing stimuli are not simply restricted to the exterior environment; they can just as well originate in an organ or other interior structure as an alert that something is amiss within. The immediate pain stimulus travels up the autonomic system to the cerebral medulla. The required response is then sent back via the autonomic nervous system. In more complex species such as humans, the signal is also communicated to the central nervous system. This leads to conscious recognition of pain and, in many cases, involvement of the higher centers of the brain. This resulting involvement of an emotional element may shape the perception of pain. Although pain plays a vital role in day-to-day function, a problem arises when the sensation persists long after the stimulus has been withdrawn. Efforts to overcome persistent pain in all likelihood trace back to prehistory. Systematic pharmacological and medical work on the relief of pain interestingly started well over a century ago, in the middle of the nineteenth century.

At its simplest level, pain is directly akin to the finger on the hot stove top. The response can be cut short at this level by numbing the fingertip. The stimulus thus never travels further up the nervous system. Towards the turn of the nineteenth century, starting with the folkloric observation of the numbing effect of cocaine, medicinal chemists developed a series of chemically related molecules that had a similar numbing action. These compounds, the local anesthetics, are still in use today for various dental and minor surgical procedures. Their local action and short duration make them unsuitable for use in prolonged pain.

Inflammation and some types of tissue injury result in the local release of a series of chemicals that cause pain. The most common manifestations of this are the aches due to arthritis and more specifically osteoarthritis. The discovery in the late 1890s that acetylsalicylic acid cured arthritis pain by the Bayer chemist Felix Hoffmann provided a treatment for that ailment. Starting in the late 1960s, that drug was supplemented by a series of new nonsteroid anti-inflammatory agents (NSAIDs). The very large number of drugs in this category are widely used for treating what may be termed moderate pain. A more detailed discussion of this class of pain-relieving drugs is given in Chapter 6.

The vast majority of pain is, however, due to major injury or disease. NSAIDS are of little use in treating this or chronic pain. Local anesthetics cover limited topical areas for relatively short periods and NSAIDs alleviate pain due to a relatively specific narrow range of causes. The drugs that have proven useful for treating deep and persistent pain probably had their origin before history came to be recorded. There is certainly no record of the person who first ingested the sap exuded on the seedpods of the poppy *Papaver somniferum* and noted that doing so made him sleepy. The dried sap, opium, became an article of commerce as early as the second millennium BC as a consequence of its soporific and narcotic activity.[1] The drug constituted one of the few items that had true physiological activity in pharmacopeias well into the nineteenth century. Abuse of the drug because of its narcotic activity led to recognition of its propensity to cause addiction. Modern research has revealed that the crude drug may contain as many as 50 compounds with very closely related chemical structures. The most active of those compounds, morphine, was first isolated from the mixture in crystalline form in 1803 by Friedrich Sertuner. The name, morphine, after the Greek god of sleep *Morpheus*, derives from what was considered the principal activity of the drug. The compound replaced opium in many applications such as analgesia because doses could be more easily and reliably gauged with the pure crystalline substance. One of the minor constituents that accompanied morphine in opium later became a drug in its own right. This compound, codeine, although somewhat less potent than its parent, is more readily absorbed on oral administration. It can be prepared in a single step from morphine as it is simply the methyl ether of the latter. A more practical route to this widely used opiate will be considered later.

From that point until about 1930, chemists in various laboratories carried out a variety of experiments to modify the structure of morphine to try to find drugs with superior activity and lower addiction potential. The chemical structure of the compound was, however, quite unknown at the time, beyond its relative empirical formula. The very concept of molecular structure was in fact not fully recognized until the end of the nineteenth century.[2] These transformations thus involved largely treating morphine with selected reagents and isolating the products, if any. The structures of these derivatives were thus not assigned until the structure of morphine itself was announced in 1925 by Robert Robinson (later Sir) and J. M. Gulland. Historical accuracy would demand that use of the structure be deferred until the chemistry had proceeded beyond that date. Authenticity will however be violated in the interest of chemical clarity.

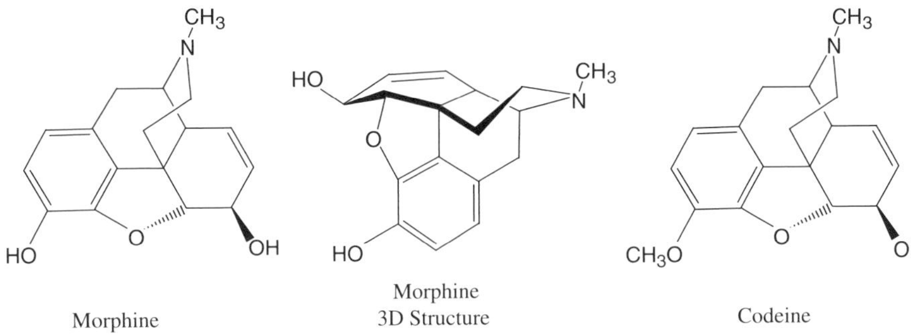

Figure 1   Morphine and codeine.

Conventional depictions of the structures of these first opiates are shown at left and right in Figure 1. The center diagram shows a three-dimensional view of morphine. Several features emerge from this picture, including the fact that the benzene ring is at a right angle to the plane of the six-membered nitrogen-containing ring. This will have an important bearing on the understanding of the mode of action of these molecules. It was established many decades later that morphine and the many structurally related molecules bind to a specific set of receptors. Those so-called opioid receptors, as is the case for most biological structures, are of course made up of molecules, mainly proteins. The selectivity of receptors for effector molecules is due to their well-defined three-dimensional structures. The proteins that make up the receptors consist of assemblages of amino acids. Each of the compounds exists in two forms whose gross formulas are identical; they do, however, differ in the fact that their three-dimensional shapes bear a mirror image relationship. Only one of those forms is used in building vertebrate proteins; in technical terms these comprise the l-amino acids. The receptor itself will then also consist of but one of two possible mirror image forms. A direct consequence of this is that receptors will bind optimally with only one of the two possible forms of a molecule, which can itself exist in two three-dimensional mirror image forms. This expectation was particularly well illustrated in the case of morphine, which, like most natural products, consists of a single stereoisomer. The unnatural mirror image form of morphine prepared by Brossi and his colleagues[3] proved to have markedly diminished binding affinity and biological activity compared to the natural isomer.

Morphine came into widespread use in the practice of medicine in the middle of the nineteenth century. This was paralleled by its abuse as a recreational drug. In addition, increased legitimate usage led to awareness of the drug's high propensity to lead to addiction on the part of long-term users. A true addict is actually physically dependent on circulating levels of drug as attested by the grim symptoms that result from sudden withdrawal. Morphine, however, was and still is an indispensable drug for treating severe acute or chronic pain. This dilemma has led to continuing research aimed at providing derivatives or synthetic compounds that act on the same receptor but do not cause the dependence that typifies morphine itself (Fig. 2). One of the first derivates to come out of such a program is a poster child for the law

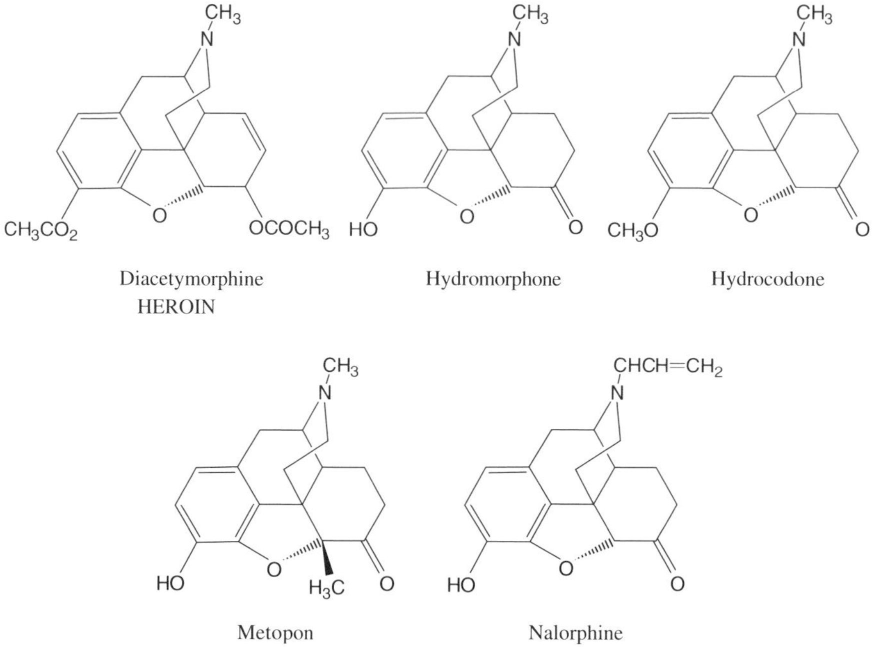

Diacetymorphine
HEROIN

Hydromorphone

Hydrocodone

Metopon

Nalorphine

Figure 2   Drugs derived from morphine.

of unintended consequences. Reaction of morphine with acetic anhydride leads to the crystalline derivative, diacetylmorphine,[4] a drug far better known as heroin. This compound is somewhat more potent than morphine itself. The lower doses that were required led to the mistaken assumption that the smaller amounts would lead to less addiction. The popularity of this drug may well trace back to the fact that it reaches the brain more quickly than does morphine on intravenous administration, that is, "mainlining." This is apparently more likely to provide the "rush" sought by addicts. Relatively simple chemical manipulations starting with morphine lead to drugs such as hydromorphone[5] and hydrocodone,[6] which are more potent than the parent molecule; the latter is often prescribed as a cough suppressant. A somewhat longer and more complex sequence leads to metopon,[7] a compound that incorporated an extra methyl group onto one of the rings. The fact that it shows substantially the same activity as morphine foreshadows the fact that biological activity of opiates is compatible with quite drastic changes in chemical structure. A substantial change in activity does, however, occur with a change of the substituent on the basic nitrogen atom. It takes a number of steps to replace the methyl group with allyl ($CH_2CH=CH_2$) to give nalorphine.[8] This compound, surprisingly, acted as an antagonist to morphine both in small animals and in several in-vitro tests at low doses. At higher doses, however, nalorphine surprisingly acted as an analgesic. Explanation of that apparently paradoxical activity awaited the discovery of the opiate receptor. In common with other such entities, the receptor was eventually found to consist of a number of subtly different subtypes, one of which interacted with nalorphine. The suggestion that nalorphine might point to analgesics with reduced addiction potential led chemists to try that same change on some later series of morphine-related compounds.

Medicinal chemists will often synthesize compounds whose structures comprise abbreviated versions of some active lead in order to better understand requirements for interaction with putative receptors. A series of opiate drugs that helped establish that minimal structure was in fact discovered adventitiously. A compound synthesized by chemists at the Sterling labs[9] showed sufficient antispasmodic activity in animal models to be taken to the clinic for that indication. The analgesic activity of the compound, meperidine (Fig. 3), once known as pethidine, was discovered in the course of those trials.[10] It quickly found a place in the clinic because of its good oral activity. More detailed work showed that the drug was a classical opiate that could be viewed pharmacologically as a version of morphine with about one-fifth the potency of the parent drug. This simplified opiate unfortunately had the same propensity to cause addiction as morphine itself. The much simpler structure, compared to morphine, led to intensive work on the preparation of related compounds in many other laboratories. It has been estimated that over 4000 analogs had been prepared by 1965.[11] Although this failed to produce the long sought after nonaddicting analgesic, it did yield a series of extremely potent analogs that will be touched on later in this account and did lead to a picture of the chemical structure required to interact with the principal opiate receptor.

Analgesic activity persists in the face of even further simplification of the structure, such as attaching the basic nitrogen on an open chain instead of a ring (Fig. 4). The impetus for this work is more difficult to establish as the work was carried out in Germany during the Second World War. The research was disclosed in a postwar intelligence report rather in the open scientific literature.[12] The drug, methadone, is an effective analgesic that interacts with the same receptor as morphine. When taken orally, the drug is however devoid of the euphoriant activity that results from morphine and more importantly heroin. The drug is thus used extensively to treat individuals who are addicted to opiates. Methadone effectively avoids withdrawal symptoms by serving as a substitute for the opiate to which they are addicted. The lack of the high allows them to carry out a normal life, although they are in fact still addicted to opiates. Injected methadone, however, has many of the same effects as drugs of abuse. Methadone clinics consequently monitor drug ingestion to prevent their clients making off with the tablet and later converting it to an

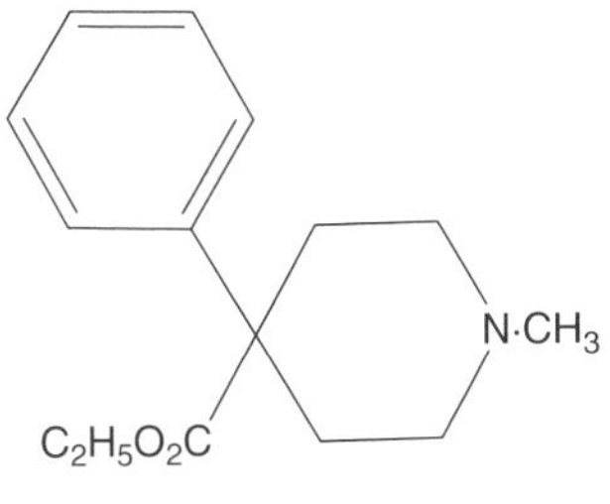

Figure 3   Meperidine.

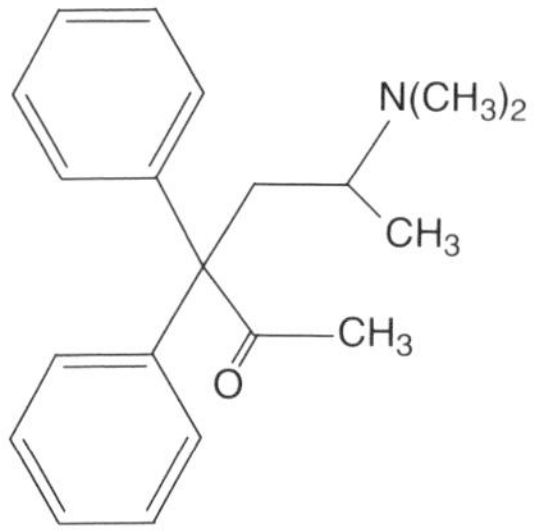

Methadone

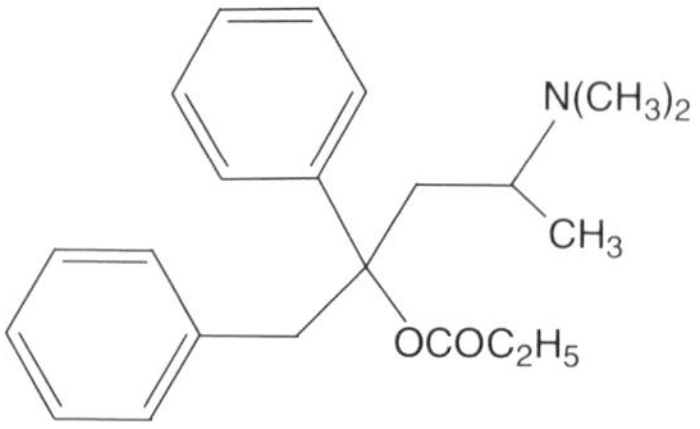

Propoxyphene

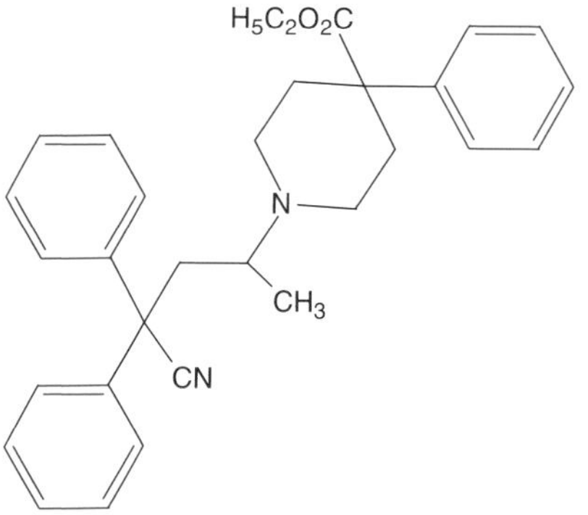

Diphenoxylate

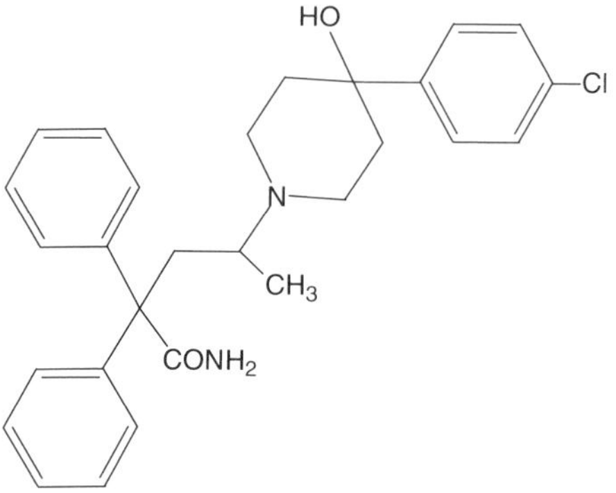

Loperamide

Figure 4 Open-chain opiates and their derivatives.

injectable form. This ultimate simplification of the opiate structure did not stimulate quite the same effort at preparing analogs as had the discovery of meperidine. Work at Lilly in the 1950s did, however, lead to the preparation of the open-chain analog propoxyphene.[13] This compound, far better known by its trade name Darvon®, has found extensive usage for treatment of mild pain. The compound does bind to opiate receptors, although not as strongly as classic opiates such as methadone. Long-term use can result in dependence, but the drug does not seem to be used recreationally.

Morphine and other opioids are associated with a set of well-known side effects. Changes in peristalsis in the GI tract that result in constipation is one of the more prominent of these effects. It has proven possible to turn what started as a shortcoming into a useful indication by manipulation of the chemical structure of methadone. By incorporating a good part of the structure of that compound, chemists at Janssen were able to devise an agent that retains the side effect, interference with peristalsis. This would in theory, as well as in practice, counteract diarrhea. The drug was acceptable for general use because of its reduced opiate activity on the central nervous system. This compound, phenoxylate,[14] found extensive use for treatment of diarrhea. Its successor from the same labs, loperamide,[15] is now familiar to world travelers as Imodium®, or one of its generic versions. Both these compounds, it should be noted, do still bind to opiate receptors.

The very complex structure of morphine proposed by Robinson and Schopf was not fully accepted for some years after it was proposed. Today's array of instrumental technology, including X-ray diffraction, nuclear magnetic resonance, and infrared absorption, makes it generally possible to assign detailed chemical structures to unknown samples with a high degree of certainty. The only proof of structure considered acceptable before the availability of those tools comprised total synthesis. This still acceptable criterion required that the compound in question be synthesized from starting materials whose structure was firmly established by a sequence of unambiguous transforms. In pursuit of that goal, Grewe in Germany succeeded in 1948 in synthesizing a compound that included a simplified form of the morphine skeleton.[16,17] This still, however, lacked the five-membered ring present in the natural product or for that matter any substitution on the terminal ring. The observation that this molecule nonetheless exhibited considerable analgesic activity in animal models spurred work in a number of laboratories aimed at investigating this basic nucleus, dubbed a morphinan (Fig. 5). Scientist at Hoffmann La-Roche published work on this series as early as 1949.[18] As general rule, organic reactions that produce molecules that can exist as mirror images will result in an exact 50 : 50 mixture of those products. The product from the Grewe synthesis thus consists of such a mixture of stereoisomers. Separation of the mixture, called

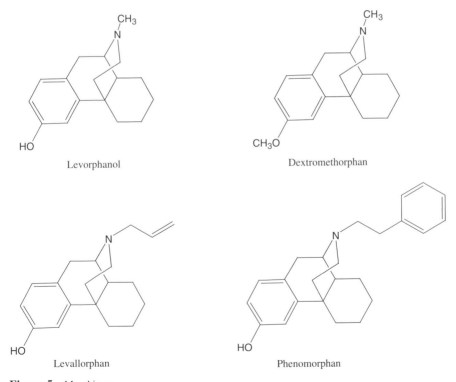

Levorphanol

Dextromethorphan

Levallorphan

Phenomorphan

Figure 5   Morphinans.

a racemate, into its individual components revealed that, as in the case of morphine, a single isomer was responsible for most of the analgesic activity.[19] Named the levo isomer for the fact that it rotates polarized light to the left, the product that consists of a single mirror image form was introduced into the clinic as levorphanol. This drug is several times more potent than morphine, although it has much the same profile of side effects and abuse potential. As in the case of the natural product, replacement of the methyl group on the nitrogen by allyl gives a compound, levallorphan, which acts as an opiate antagonist. The compound in which the double bond in allyl forms part of a benzene ring, phenomorphan is a very potent analgesic with no antagonist activity. Neither of these have been approved for use, although the latter is found on lists of controlled substances. Isolation of levorphanol from the mixture leaves behind the other mirror image form. This compound, however, retains the cough suppressing activity found in many opiates. Conversion of that compound to its methyl ether affords dextromethorphan, a compound found in many over-the-counter cough and cold remedies.

The next step towards yet another simplification of the once complex structure can be attributed to advances in organic synthesis methodology. The system that lacked yet another ring, called a benzazocine, had been originally prepared by a somewhat complex scheme. Application of the chemistry pioneered by Grewe led to its accessibility in a relatively few steps.[20] The simplest product from this sequence, whose name is here simplified to N-methylbenzocine, proved to have a biological profile quite similar to that of morphine with potency in the same range (Fig. 6). The ready availability of this compound led many laboratories to explore the series in the search for the ever elusive nonaddicting analgesic. The group at Smith Kline French, for example, explored the N-allyl derivative. This compound proved to be an opiate antagonist like its counterpart in the morphine series, nalorphine. Unlike its counterpart, however, it was devoid of analgesic activity.[21] The phenylethyl susbstituted derivative phenazocine, as in the benzomorphan series, proved to be a very active analgesic with, in this case, about ten times the potency of morphine.[22] The compound that showed a somewhat different profile came from the laboratories of Sterling Winthrop.[23] This derivative may be viewed as having a couple of additional methyl groups attached to the end of the allyl side chain on the nitrogen. This agent, later named pentazocine, showed a mixture of agonist and antagonist activities. In the clinic this drug, soon trademarked as Talwin®, proved to be an effective analgesic with less abuse potential than the classic opiates. For several decades, in fact, this drug was not included on DEA lists of scheduled drugs. It is now included, although in the least stringent category IV. A set of further modifications led to ketazocine, yet another drug with mixed agonist–antagonist activity. The current view that opiate receptors exist as several distinct forms may have been initially prompted by the need to account for the spectrum of biological activities of this compound and of pentazocine. The existence of the multiplicity of receptors has been borne out by a host of receptor-binding studies using many different synthetic analgesics as well as the endogenous substances that were soon discovered. To admittedly grossly oversimplify, morphine, its very direct derivates, and many purely synthetic compounds such as meperidine, bind

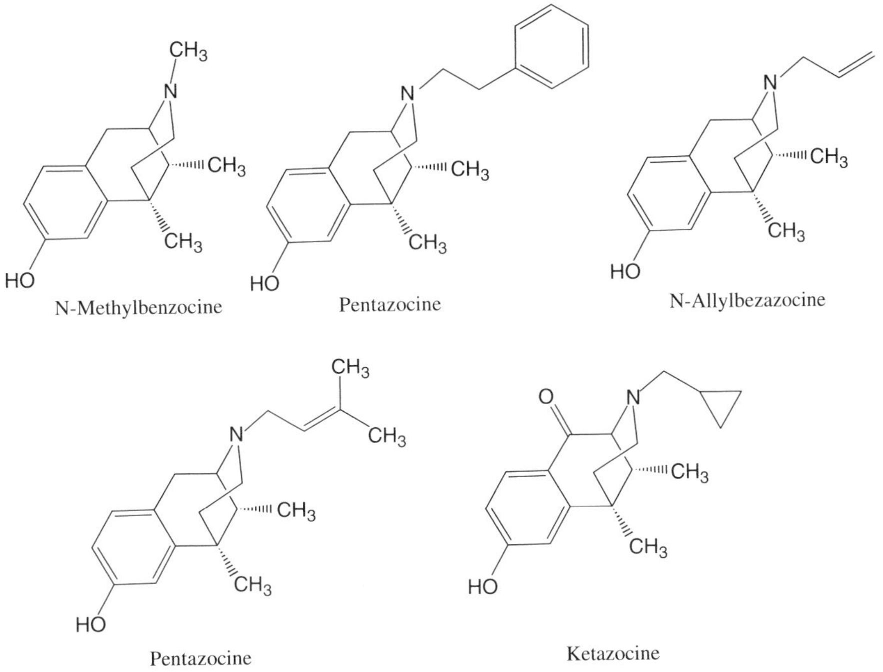

Figure 6 Benzazocine analgesics.

to a receptor designated as mu (μ). Compounds in this class will, it is posited, show good analgesic activity, but will also have the same shortcomings as the prototype. Compounds such as ketazocine act mainly on a receptor designated kappa (κ); these drugs are also analgesic, but tend to act as antagonists at the mu receptor. Pentazocine to some extent acts at both sites. Several other subtypes have been identified in recent years, including delta and sigma receptors. Subtypes of each have been identified as well. Details of that work are well beyond the scope of this discussion.

Analogs based on the full morphine structure had not in the meantime been neglected. A major program aimed at non- or at least less-addicting centrally acting analgesics was based on the discovery of the potentiating effect of including an additional oxygen atom in the molecule. The poppy *Papaver somniferum*, as noted above, produces a host of very closely related alkaloids (Fig. 7). One of those, thebaine, has little activity in its own right but proves to be a very useful starting material for producing modified structures. An additional hydroxyl group is added to the molecule, for example, when the compound is oxidized with hydrogen peroxide. The conventional depiction seems to indicate that this new function occupies a very crowded position. The three-dimensional view in Figure 7 show that the newly introduced function in fact resides on the more open face of the molecule. This first product is readily transformed in several steps to hydroxymorphone, one of the most potent analgesics based on the morphine struture.[24] A different set of transformations lead to oxycodone; this drug is a good bit more potent than codeine,

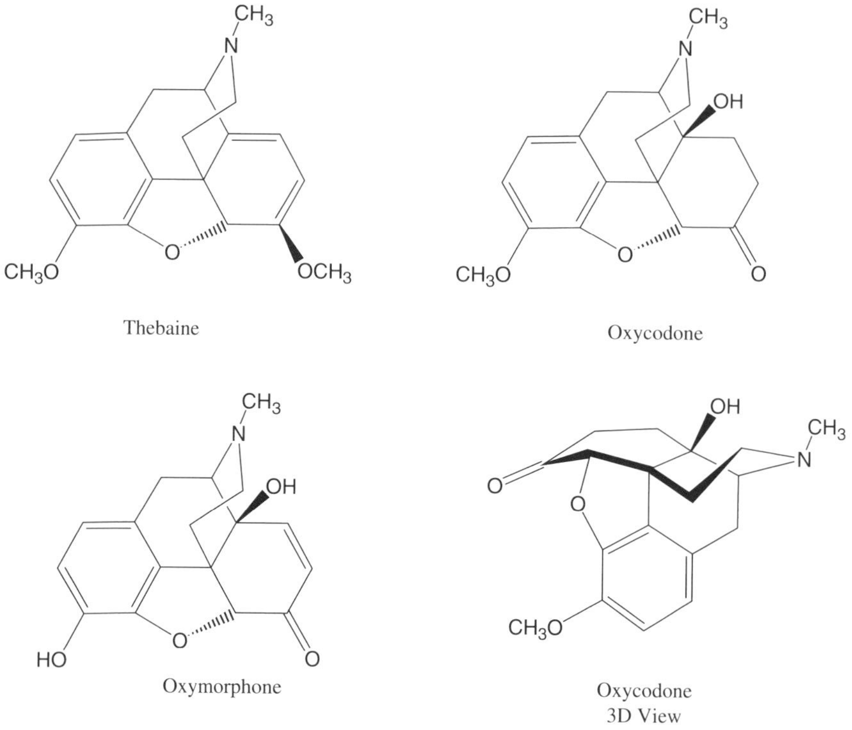

Thebaine

Oxycodone

Oxymorphone

Oxycodone
3D View

Figure 7 Thebaine-derived opiates.

its counterpart lacking the extra oxygen atom. This last compound has recently gained much press coverage under the trade name OxyContin®. In fact it is simply a controlled-release tablet intended for treatment of chronic pain. Abusers apparently extract the active drug substance from the tablet. Oxycodone itself is probably no more addicting than any other related opiates, popular and regulatory impressions notwithstanding.

The growing importance of drugs based on thebaine led to a search for an alternative supply of this alkaloid, because it is only a minor constituent of the mixture from the opium poppy. Several such strains have since been identified, including *Papaver bracteum*, whose dried sap is reported to contain as much as 26% thebaine.[25] Most of the codeine used today is probably produced from thebaine rather than derived from morphine.

Replacing the methyl group on the nitrogen by an allyl in this series gives naloxone,[26] a compound that exhibits potent antagonist activity at opiate sites (Fig. 8). The drug shows no analgesic activity in any test systems, in contrast to the allyl derivatives of the morphine, benzomorphan, or for that matter benzazocine series. Administration of this potent antagonist quickly reverses the actions of opiates; it will in consequence precipitate withdrawal symptoms when administered to addicted individuals. Naloxone is, however, not active when taken by mouth. It had thus been proposed to combine the drug with methadone. Individuals in

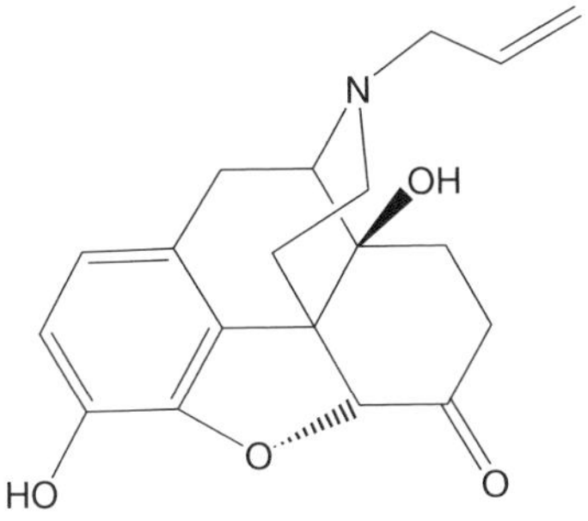

Naloxone

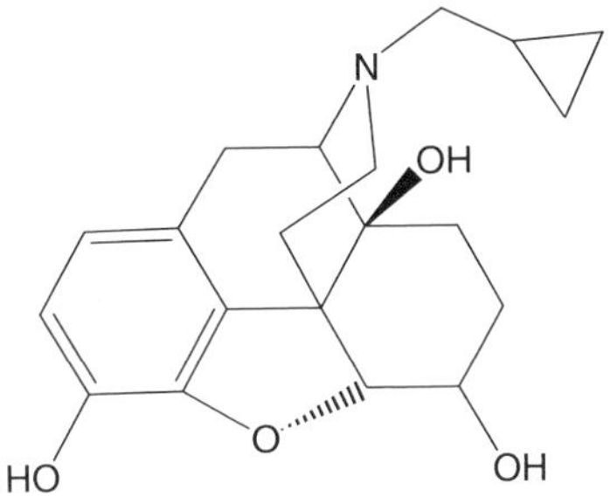

Nalbufine

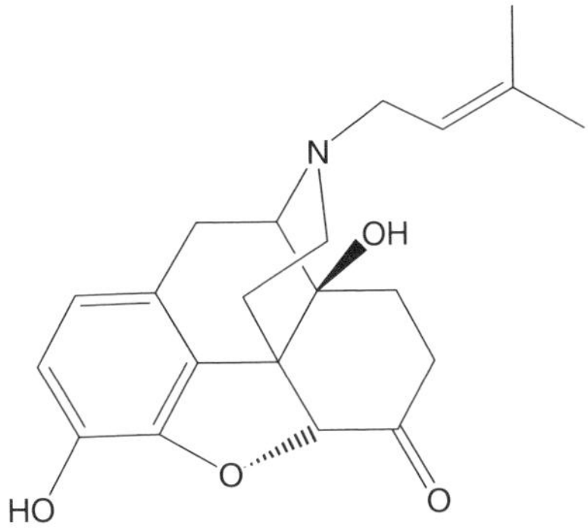

Naltrexone

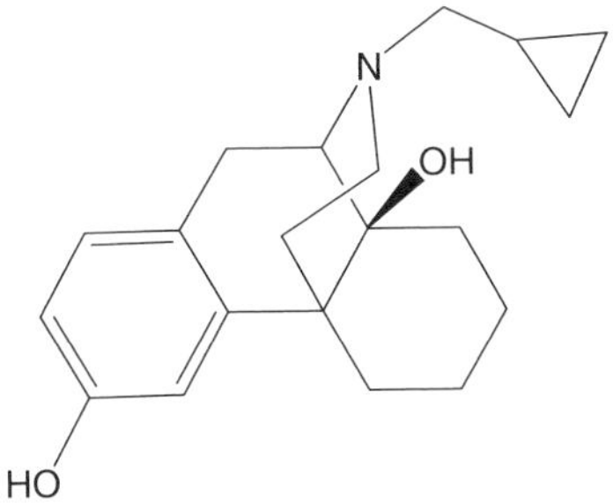

Butorphanol

Figure 8   Opiate mixed agonists–antagonists.

maintenance programs would obtain the full benefit of methadone as long as they took their dose orally. Converting the drug to an injection would be self-defeating as the naloxone would completely abolish the activity of methadone. In the benzazocine series, replacing methyl by an allyl group with two methyl groups led to pentazocine, an analgesic with much reduced abuse potential. The same change in the present series lead to naltrexone. Antagonist activity seems to prevail with this drug. It is, however, approved for maintenance treatment of alcoholism and to a lesser extent opiate addiction. In the meantime, analgesic activity predominates with nalbufine, in which one of the hydrogens on the methyl on the nitrogen is replaced by a three-membered cyclopropyl ring.[27] This compound is approved as an opiate analgesic. The structure of butorphanol is a hybrid between the morphinan and thebaine derived series in that the compound includes the extra hydroxyl group found in the latter but lacks the fused oxygen-containing ring. The agent is thus produced by total synthesis[28]; the commercial form interestingly consists of but one of the possible stereoisomers. The drug is indicated for pain relief as analgesic activity predominates in this mixed agonist–antagonist. It is usually administered by injection as it is quickly metabolized when taken by mouth. It has been recently introduced as a nasal spray to take advantage of the ready absorption of drugs by the nasal mucosa.

The rich chemical functionality available in the structure of thebaine led chemists at Reckitt and Coleman to investigate compounds produced by adding yet another bridging ring to the basic molecule. This transform was made possible by the existence to the two sequential double bonds in the right-hand ring.[29] The class of resulting compounds was given the trivial name oripavanes (Fig. 9). Several further transforms on the product led to etorphine, one of the most potent known opiates. This agent shows 1 : 10,000 higher potency than morphine in various animal models. The drug has little clinical use, but has been used extensively to knock down large animals in captivity or in the wild preparatory to various procedures. Its very potency poses a real hazard to those who handle the compound; accidental ingestion or injection can quickly lead to loss of consciousness and respiratory arrest. To mitigate the danger, etorphine kits are thus supposed to always include a syringe filled with naloxone.

Further chemical transforms that have led to mixed action compounds in other series have led to corresponding changes in activity among oripavanes. The drug

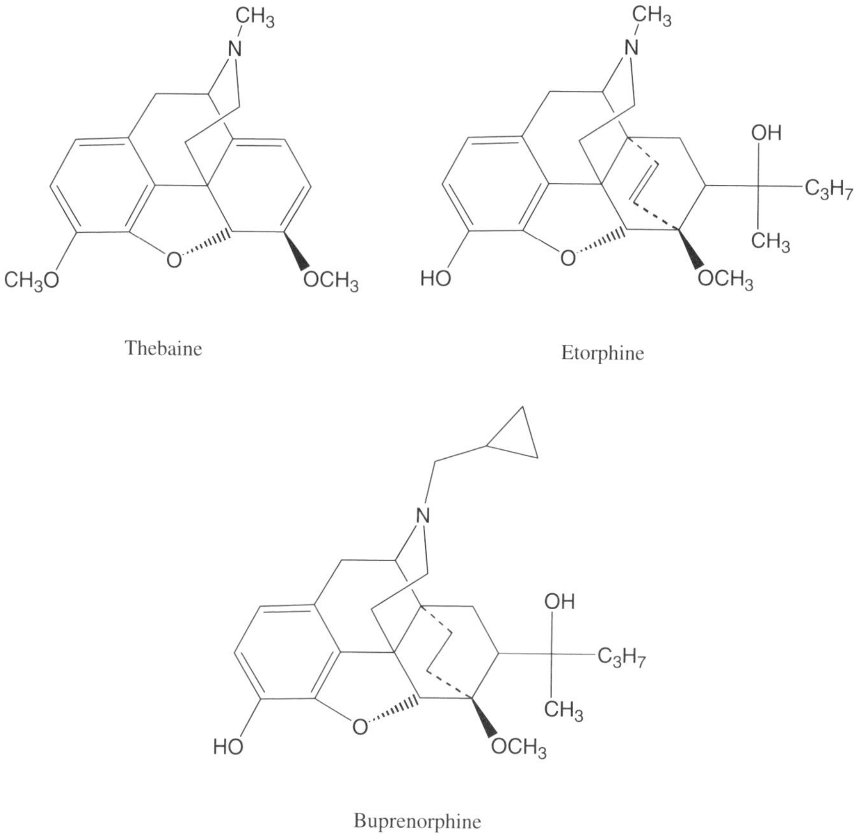

Thebaine

Etorphine

Buprenorphine

Figure 9   Oripavanes.

that incorporates a cyclopropyl ring, buprenorphine,[30] is a mixed agonist–antagonist that substitutes for opioids in addicts. It does, however, produce only a limited amount of euphoria. It has, in consequence, been investigated extensively as an alternative to methadone for treatment of opiate addicts. As is the case of the latter, co-administered naloxone will completely wipe out the drug effect. A tablet for support of addicted individuals that combines buprenorphine and naloxone was approved by the FDA in 2002.

The forgoing discussions of opiate receptors have sidestepped the reason for the very existence of those entities. The receptors for catecholamines discussed in Chapter 3 clearly exist to mediate neurotransmitter response. It is highly unlikely that compounds related to morphine play a similar role in vertebrates given that most lead a quite successful life without ever seeing any of those molecules. The development by the late 1960s of ever more sensitive methods for studying receptors and the substances that they bind made it possible to tackle that question. The isolation by Pert and Snyder of opiate receptors from nerve tissue[31] provided the necessary tool for studying those interactions. Within the next few years, these same investigators identified a morphine-like peptide in bovine brain that elicited analgesia in rats.[32] Purified preparations of this class of substance, later named endorphins, were found to bind to the same receptors as morphine itself. The endorphins comprise a group of related peptides that include from 16 to 30 amino acids. Although their exact role is still open to discussion, they act as both neurotransmitters and neurohormones, and tend to modulate the pain response. More detailed later work in this area identified a pair of closely related peptides called enkephalins, which also bind to opiate receptors. These much smaller peptides are composed of but five amino acids and differ only by the identity of a single amino acid at one end. The enkephalin compounds, which are found only in the brain, act as true opiates in various animal models.[33] A recently isolated peptide, morphiceptin, which comprises but four amino acids, has been described as one of the most selective agonists for mu opiate receptors.[34]

The question of the influence of stereoisomerism on the activity of the structurally quite complex compounds more directly related to morphine is moot, because the starting materials occur as a single one of the two possible mirror image forms. Products derived from those compounds will also as a rule comprise a single stereoisomer. This question is also not relevant in the much simpler drug meperidine (Fig. 3) because of the lack of any asymmetric elements in the structure. Adding a single methyl group to the ring adds that missing element. The resulting molecule now exists as a pair of mirror image forms and the two possible isomers can no longer be superimposed. The group at Lilly synthesized the meperidine-related compound, picenadol, which included the extra ring methyl group;[35] in addition, the ester group was replaced by a propyl group and a phenolic hydroxyl function was added to the benzene ring (Fig. 10). That last change led to increased potency in the simpler meperidine series. Separating the resulting stereoisomers, they found that one compound acted mainly as an opiate agonist and the other showed mostly antagonist activity. The mixture, picenadol, showed the properties of a mixed agonist–antagonist. Some years later, taking the clue from this finding, they modified the molecule so as to increase its affinity for water, a change that should prevent

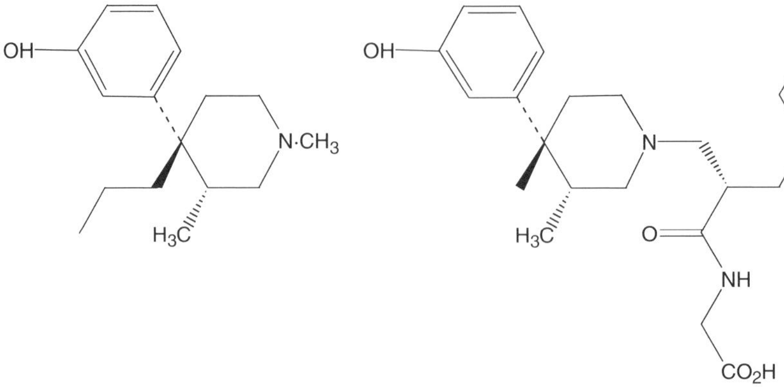

Picenadol   Alvimopam

Figure 10   Recent compounds related to meperidine.

it from reaching the brain. The mirror image form of that new compound, which corresponds to the antagonist isomer of picenadol, retained that activity.[36] The resulting product, alvimopam, is an opiate antagonist that acts largely on opiate receptors in the intestinal tract. The drug thus counteracts the effects of opiates on the gut. It has proven useful in treating the paralysis of the lower GI that often follows major surgery (ileus) or that produced by opiates. The compound interestingly does not interfere with opiate analgesia.[37]

All the compounds considered to this point, be they agonists or antagonists, contain a common set of structural elements. That core was recognized back in 1959 in the proposal by Beckett and Casey.[38] They found that this comprised a benzene ring attached to a carbon that had two other substituents, neither of which was hydrogen, a so-called quaternary center; a basic nitrogen atom was located at a remove of two carbon atoms from the same center. It should be added, as an aside, that such generalizations tend to have the element of self-fulfilling prophecies. It is a rare chemist who will set out to deliberately prepare compounds with random structures; the great majority will choose compounds known to have activity as the model for their projects. Those additional examples, if active, will reinforce the generalization.

The structure of the relatively simple compound tramadol[39] does not bear a close relation to either the meperidine or the morphine-derived series (Fig. 11). It does, however, incorporate the elements of the Beckett and Casey rule: benzene at a quaternary center and nitrogen at a two-carbon-atom remove. This drug seems to be gaining considerable attention, and, more importantly, usage. Although, the agent binds to opiate receptors, its pharmacological profile differs in ways that offer reduced addiction liability. This is reflected in the fact that the drug is listed in the least closely regulated category: Schedule IV. The structure of tramadol does, however, bear an interesting relation to the two-ring opiate tilidine.[40] This older compound is a classical opioid analgesic. The structure of this compound, however, does not fit the rule, as nitrogen is but a single atom from the quaternary

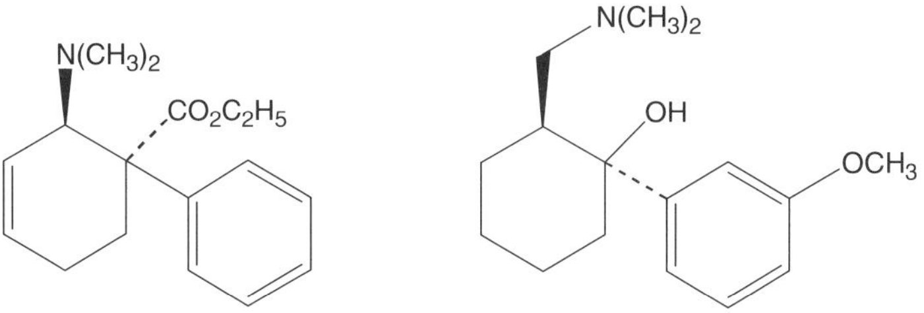

Tilidine   Tramadol

Figure 11 Tilidine and tramadol.

center. When first introduced, tilidine was believed to offer reduced abuse liability. This expectation was dispelled as the drug came into widespread use. This is reflected in the fact that this is a more closely regulated Schedule II agent.

The structure of a series of compounds from the Janssen labs also pointed to the shortcomings of the Beckett and Casey rule. The first of these, fentanyl,[41] is an extremely potent analgesic that binds with high affinity to mu opioid receptors (Fig. 12). The compound is 50 times more potent than morphine in humans and as much as 100 times more potent in experimental animals. Some later analogs from competing labs, such as brifentanil,[42] showed that there was considerable latitude in the structural modifications compatible with activity. This particular agent apparently offered too few advantages over existing compounds to merit clinical use. A compound that showed 12,000 times the potency of morphine synthesized by chemists at Upjohn departed even more radically from the Beckett and Casey postulate. The observation that a model of that compound could provide a four-point overlay on a model of fentanyl led them to propose a modification of the rule that posited a fourth potency-enhancing binding site.[43]

Adding a carbon atom in the fentanyl series to the position two atoms removed from nitrogen in effect restores the quaternary center required by the Beckett and Casey postulate. Two of the products from this modification, alfentanil and sufentanil,[44] are potent opiate analgesics currently approved for use in the clinic (Fig. 13). Another, carfentanil,[45] is so potent that its use, like etorphine, is restricted to

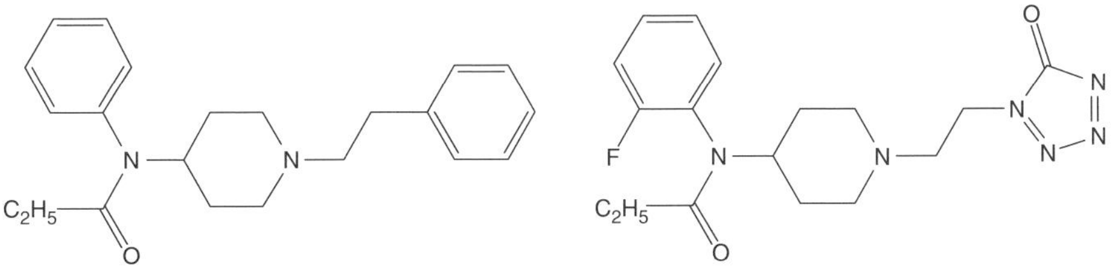

Fentanyl   Brifentanyl

Figure 12 Fentanyls.

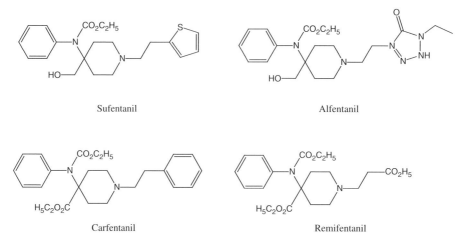

Sufentanil

Alfentanil

Carfentanil

Remifentanil

Figure 13   More fentanyl derivatives.

knocking down large animals. The dose required to knock down a moose ranges from 6 to 14 μg per kilogram. This amounts to about 6 mg total drug per animal.

Much work has been devoted to developing controlled release forms of drugs that will provide sustained blood levels over prolonged periods. This is best exemplified in the case of central analgesics by OxyContin®. Some attention has recently been directed to drugs that are quickly inactivated. In the case of analgesics used in surgery, this stratagem will allow a drug to be cleared as soon as the procedure is completed, presumably diminishing side effects from prolonged exposure to an opiate. This concept has been applied by attaching an ester side chain to fentanyl; the benzene ring at the end of the chain is in this case replaced by a carboxylic ester. Once in the bloodstream after intravenous administration, the terminal ester in remifentanil[46] is hydrolysed by the plentiful serum esterases. The resulting very polar acid can now no longer penetrate the blood–brain barrier. This form too can be more readily excreted than a neutral opiate.

REFERENCES

1. A. G. GIBSON, *Plants and Civilization*, available at http://www.botgard.ucla.edu/html/botanytextbooks/economicbotany/Papaver/
2. See J. BUCKINGHAM, *Chasing the Molecule*, Sutton Publishing, Gloucestershire, UK, 2004.
3. IIJIMA, J.-I. MINAMIKAWA, K. C. RICE, A. E. JACOBSON AND A. BROSSI, *J. Org. Chem.*, 43, 1462 (1978).
4. O. HESSE, *Ann.*, 220, 203 (1883).
5. H. RAPPOPORT ET AL., *J. Org. Chem.*, 15, 1103 (1950).
6. N. B. EDDY AND J. G. REID, *J. Pharmacol. Exp. Therap.*, 52, 468 (1934).
7. L. SMALL, H. M. FITCH AND W. E. SMITH, *J. Am. Chem. Soc.*, 58, 1457 (1936).
8. J. WEIJLARD AND A. E. ERICKSON, *J. Am. Chem. Soc.*, 69, 869 (1942).
9. O. EISLEB, U.S. patent 2,167,351 (1939).

10. O. Eisleb and O. Schaumann, *Deutsch. Med. Wochenschr.*, 65, 967 (1939).

11. L. S. Harris, *Annu. Reports. Med. Chem.*, 1, 40 (1965).

12. E. C. Kleiderer, J. B. Rice, V. Conquest and J. H. Williams, Report PP-981, Office of the Publication Board, Dept. of Commerce, Washington, DC, 1945.

13. R. A. Pohland and H. R. Sullivan, *J. Am. Chem. Soc.*, 77, 4458 (1955).

14. P. A. J. Janssen, A. H. Jagenau and J. Huygens, *J. Med. Chem.*, 1, 299 (1959).

15. R. A. Stokbroekx, J. Vandenbenk, A. H. M. T. vanHeertum, G. M. L. W. Van Laar, M. J. M. C. van der Aa, W. F. M. vanBeeren and P. A. J. Janssen, *J. Med. Chem.*, 16, 782 (1973).

16. R. Grewe and A. Mondon, *Chem. Ber.*, 81, 279 (1948).

17. For a review of early work see D. Lednicer, *J. Chem. Ed.*, 66, 718 (1989).

18. O. Schnnider and A. Grussner, *Helv. Chim. Acta.*, 32, 821 (1949).

19. O. Schnnider and A. Grussner, *Helv. Chim. Acta.*, 34, 2211 (1951).

20. For a review see D. C. Palmer and M. J. Strauss, *Chem. Rev.*, 77, 1 (1977).

21. M. Gordon, J. J. Lafferty, D. H. Tedeschi, N. B. Eddy and E. L. May, *Nature*, 192, 1089 (1962).

22. E. L. May and N. B. Eddy, *J. Org. Chem.*, 24, 295 (1959).

23. S. Archer, N. F. Albertson, L. S. Harris, A. K. Pierson and J. G. Bird, *J. Med. Chem.*, 7, 123 (1964).

24. U. Weiss, *J. Am. Chem. Soc.*, 77, 5891 (1955).

25. N. Shargi and L. Lalezari, *Nature*, 213, 1244 (1967).

26. M. Lewenstein and J. Fishman, U.S. patent 3,254,088 (1966).

27. I. J. Pachter and Z. Matossian, U.S. patent 3,393,197 (1968).

28. I. Monkovic, T. T. Conway, H. Wong, Y. G. Perron, I. J. Pachter and B. Belleau, *J. Am. Chem. Soc.*, 53, 7910 (1973).

29. K. W. Bentley, D. G. Hardy and B. Meek, *J. Am. Chem. Soc.*, 87, 3273 (1967).

30. K. W. Bentley, U.S. patent 3,433,791 (1969).

31. C. B. Pert and S. H. Snyder, *Science*, 179, 1011 (1973).

32. C. B. Pert, R. Simantov and S. H. Snyder, *Brain Res.*, 18, 523 (1977).

33. For a discussion of synthetic enkephalins see J. S. Morley in D. Lednicer, Ed., *Central Analgesics*, Wiley, NY, 1982, pp. 81–135.

34. A. Janecka, J. Fichna, M. Miorwski and T. Janecki, *Mini. Rev. Med. Chem.*, 2, 565 (2002).

35. D. M. Zimmerman, S. E. Smits, M. D. Hynes, B. E. Cantrell, J. D. Leander, L. G. Mendelson and R. Nickander, *Drug Alcohol Depend.*, 14, 381 (1985).

36. D. M. Zimmerman, J. S. Gidda, B. E. Cantrell, D. D. Schoepf, B. G. Johnson and J. D. Leander, *J. Med. Chem.*, 37, 2262 (1994).

37. P. Neary and C. P. Delaney, *Expert. Opin. Investig. Drugs*, 14, 479 (2005).

38. A. H. Beckett and A. F. Casey, *J. Pharm. Pharmacol.*, 6, 986 (1959).

39. K. Flick and E. Frankus, U.S. patent 3,652,589 (1972).

40. G. Satzinger, *Ann.*, 738, 64 (1962).

41. W. F. M. VanBeever, C. J. E. Niemeegers and P. A. J. Janssen, *Arnzeim. Forsch.*, 13, 502 (1963).

42. N. Lalinde, J. Moliterni, D. Wright, H. K. Spencer, M. H. Ossipov, T. C. Spaulding and F. G. Rude, *J. Med. Chem.*, 33, 2876 (1990).

43. See D. Lednicer, in D. Lednicer, Ed., *Central Analgesics*, Wiley, NY, 1982, pp. 203–206.

44. W. F. M. vanBeever, C. J. E. Niemeegers, K. H. Schellekens and P. A. J. Janssen, *Arnzeim. Forsch.*, 26, 1584 (1976).

45. P. G. H. vanDaele, M. L. F. DeBruyn, J. M. Boey, S. Sanczuk, J. T. M. Agten and P. A. J. Janssen, *Arzneim. Forsch.*, 26 (1976).

46. P. L. Feldman, M. K. James, M. C. Brackeen, F. Marcus, J. Billota, S. V. Scuster, A. P. Lahey, M. R. Johnson and H. J. Leighton, *J. Med. Chem.*, 34, 2202 (1991).

Chapter 6

Nonopiate Analgesic Agents

A wide array of drugs is currently available for relief of pain caused by arthritis, inflammation, or minor injury. These same agents are also widely used to treat passing everyday aches and pains such as headaches, toothaches, and in the words of advertisements, neuritis and neuralgia. These compounds are distinct from the opiates in that they exert their analgesic action at the site of the injury that leads to pain. They are, as a consequence, sometimes called peripheral analgesics to differentiate the class from the central analgesics, discussed in Chapter 5, that act by way of the central nervous system. As more drug entities became available, their ability to reduce inflammation came to the fore. The corticosteroids to be discussed later are still the treatment par excellence for treating serious acute inflammation. It had become clear by the middle 1960s that the ancillary effects of the steroids limited their long-term use. The nonopiate analgesics provided a method for long-term treatment of inflammation. As a result, these drugs are now quite often called non steroid anti-inflammatory agents or NSAIDs.

Peripheral analgesics comprise some of the earliest examples of therapeutic agents produced by chemical synthesis. They owe much of their discovery to the growth of the synthetic dye industry in Germany towards the second half of the nineteenth century. The large-scale chemical reactions carried out by the dye industry led to the search for other applications of the new technology. It soon became a matter of course to test the activity of some of the simpler products produced by organic chemists. Although aspirin came to be the best-known example in this group, its use as a drug was in fact preceded by several other compounds (Fig. 1). Its chemical precursor, salicylic acid, was introduced in 1875 as an antipyretic for the relief of high fevers.[1] Use as an analgesic occurred within a short time. This was followed some years later by the somewhat more structurally complex compound antipyrine, its name indicating its imputed activity for reducing fevers. That drug too found use as an analgesic. It is still, interestingly, in use as a topical treatment in combination with the local anesthetic benzocaine for treating the swelling due to ear infections. The use of several other early compounds also persisted well into the present era.

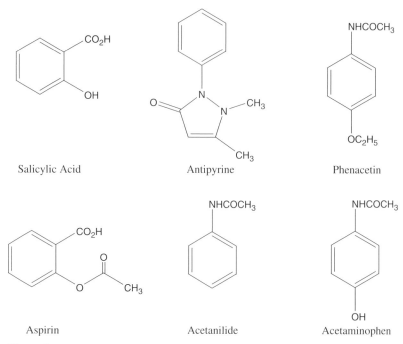

Figure 1   Early nonopiate analgesics.

The drug phenacetin was widely used in headache remedies up to the mid-1980s, at which time it was removed from the market as a result of adverse findings in animal carcinogenicity studies. Although the chemical that became Tylenol®, acetaminophen, originated in the same early period, its peripheral analgesic activity was not discovered until well into the twentieth century as a result of the search for a less toxic replacement for acetanilide, another early nonopiate analgesic. It should be noted in passing that phenacetin, acetanilide, and acetaminophen belong to subclass of peripheral analgesics that are devoid of anti-inflammatory activity. The folkloric usage of the natural product salicilin from willow bark as an analgesic and treatment for arthritis has often been credited as the inspiration that led the discovery of salicylic acid, an account now questioned by some historians. Whatever the origin of its discovery, the salicylic acid was in fairly widespread use towards the end of the nineteenth century. The unpleasant taste of the acid and its propensity to irritate the stomach led to the search for a better tolerated compound. A commonly accepted account credits Felix Hoffmann, a chemist at Bayer in Germany, with the discovery. According to the history found on the official Bayer website, Hoffmann was looking for a better alternative for the salicylic acid his father used for treating his arthritis. The acetyl derivative, which had been synthesized several decades earlier, proved to be a better tolerated but equally effective drug. It has since become a compound that is produced in tonnage amounts. Most intriguing is the fact that new activities and uses for the venerable aspirin turn up with every passing year.

The less than satisfactory pain relief often afforded by aspirin and related drugs led to continuing research aimed at producing more effective drugs. In spite of their limitations, existing drugs were nonetheless used quite widely. This revealed the fact that long-term use of these drugs caused stomach ulcers in a significant proportion of users. Finding compounds free of that side effect thus became an additional target for research on this class of drugs. Literally hundreds of peripheral analgesics have subsequently advanced far enough to be assigned nonproprietary names.[2] Well over a third of those have been approved by the U.S. FDA.

The decade after the end of the Second World War, as noted in previous chapters, was a particularly fruitful period for drug discovery. This is the time in which the thiazide diuretics, several sulfa drugs, and some of the synthetic opiates, for example, first emerged from the laboratory. This period too saw the discovery of what proved to be a new classes of peripheral analgesics. One approach to the creation of such novel agents took the structure of salicylic acid as the starting point. Replacement of the hydroxyl group in that molecule with nitrogen attached to a substituted benzene ring led to a compound whose structure is in a class with the trivial name anthranilic acid (Fig. 2). This product had much the same activity as aspirin . The nitrogen atom in this structure is barely basic because of its attachment to two benzene rings. This drug, flufenamic acid,[3] is still on the market and indicated for treating pain and inflammation. The congener with the somewhat different substitution on one of the rings, mefenamic acid, has the same activity as its predecessor.[4] Peripheral analgesic and anti-inflammatory activity is retained even when the carboxylic acid group is attached to a nitrogen-containing pyridine ring, as in flunixin.[5]

The chemical structure of the first compound in a quite different class, phenyl-butazone,[6] is sufficiently different from the other drugs then available to make it likely that it was discovered in a random screen for anti-inflammatory agents (Fig. 3). The chemical structure of the compound lacks the carboxylic acid group (CO_2H) that would be present in aspirin, the fenamates, and many of the later compounds. The very acidic proton supplied by the carboxyl group is, however, present in surrogate form on the carbon between the two carbonyls ($C=O$).

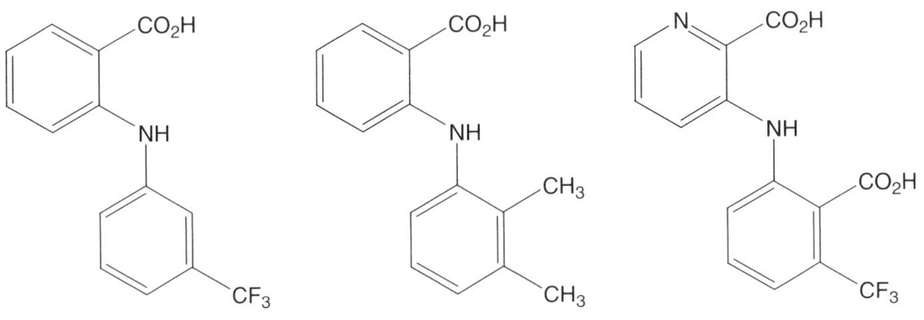

Flufenamic Acid   Mefenamic Acid   Flunixin

Figure 2 Anthranilic acid analgesics.

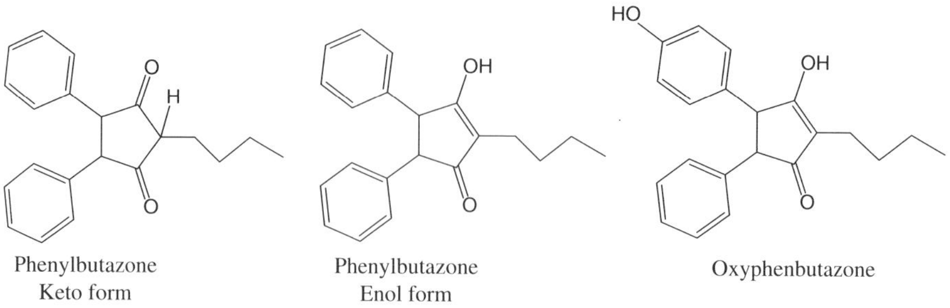

Phenylbutazone
Keto form

Phenylbutazone
Enol form

Oxyphenbutazone

Figure 3 Phenylbutazone and its metabolite.

Compounds with this function readily undergo intraconversion between two forms called tautomers. One of those has two carbonyls; the alternative isomer formed by a simple shift of a double bond places the acidic proton on oxygen and is called the enol isomer. A similar functional group will be seen later in a several much more recent drugs. Phenylbutazone is oxidized in the liver to the corresponding hydroxylated derivative. That product, oxyphenbutazone,[7] is a drug in its own right in synthetic form. The several other close analogs that have been assigned nonproprietary names are of minor importance as they offer no advantages over the original drug. Phenylbutazone itself finds considerable use in veterinary medicine, particularly in treating arthritis in horses. Its use is racehorses is, however, grounds for disqualification.

Good anti-inflammatory activity is retained when an extra carbon is inserted between the benzene ring and the carboxylic acid. This changes the side chain to an acetic acid. The sizeable number of structurally quite diverse compounds in this class that have been assigned nonproprietary names reflects the fact that the seemingly small change provided a rich source for novel compounds. It also illustrated the relatively loose structural constraints on NSAID activity. The anti-inflammatory drug diclofenac,[8] on which a benzene ring bears the acetic acid side chain, is better known as the widely used Voltaren® (Fig. 4). The extra benzene ring, which is connected by a nitrogen bridge in diclofenac, can alternatively be attached via an oxygen bridge as in fenclofenac.[9] Although this related drug showed reasonable

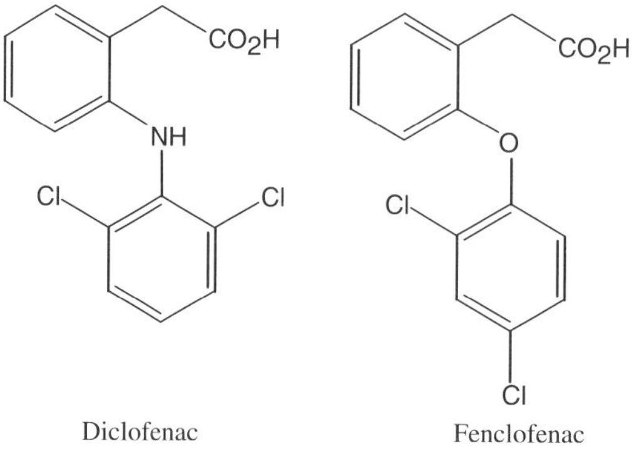

Diclofenac

Fenclofenac

Figure 4 Acetic Acid NSAIDs based on benzene rings.

activity in humans, it was withdrawn as a drug because of the large number of adverse effects that emerged after it had been in use for some time. A highly simplified compound in the acetic acid series, ibufenac, also showed very promising activity as an anti-inflammatory agent in clinical trials. It too had to be withdrawn, in this case because of liver toxicity, which showed up on prolonged use.

The rationale that led to the next series of NSAIDs was due in good part to the fact that the process of inflammation was not well understood in the middle to late 1960s when much of this work took place. There was at that time some evidence that pointed to the involvement of compounds such as serotonin in the inflammatory response. The structure of this compound, which is in fact involved in a variety of biochemical processes, comprises a modified indole, that is, a benzene ring fused to a nitrogen-containing, five-membered ring. Based on this premise, chemists at Merck prepared a series of compounds that incorporated a serotonin-like fragment in the hope of finding an antagonist that fit a putative serotonin site involved in inflammation. They were rewarded by finding that an analog from that program had outstanding anti-inflammatory activity in their animal models. This compound, indomethacin,[10] went on to become a very successful and widely used drug (Fig. 5). The mechanism by which it reduced inflammation, it turned out later, had little if anything to do with serotonin. Continued research that involved further modification of the structure revealed that good activity was maintained in a related compound in which the nitrogen in the five-membered ring was replaced by carbon. That compound, sulindac,[11] also became a very successful drug. More detailed pharmacology showed that most of the activity of this compound was in fact due to a metabolite rather than sulindac itself. The number of such cases that have already been discussed almost invariably involved oxidation of the first administered compound. In this case, however, the more potent species is generated by a reduction reaction that removes oxygen.[12]

The ensuing years saw the introduction of a very large number of NSAIDs that included acetic acid side chains. These drugs all showed similar activity as anti-inflammatory agents, although they varied in potency and duration of action. The pattern of side effects, particularly the propensity to cause stomach ulcers, was also quite similar. One of the first series of follow-on compounds came from the finding from the laboratories of MacNeill that the benzene ring fused onto the five-membered ring in indomethacin could be omitted (Fig. 6). The acetic acid side chain in both tolmetin[13] and its close relative zomepirac[14] is attached directly to an isolated five-membered, nitrogen-containing pyrrole ring, a so-called heterocycle. The benzene ring is still present, but is no longer fused to the five-membered ring. The NSAID clopirac,[15] which is available in Europe, represents a further simplification yet. The acetic acid side chain in yet another NSAID, fentiazac,[16] is attached to a five-membered thiazole ring that includes sulfur in addition to nitrogen.

Chemists in the laboratories of the Boots company where ibufenac had originated continued to prepare additional derivatives of that compound spurred by the very promising activity of this drug.[17] What again looks like a very minor change led not only to a very successful drug, but also to a class of compounds that all

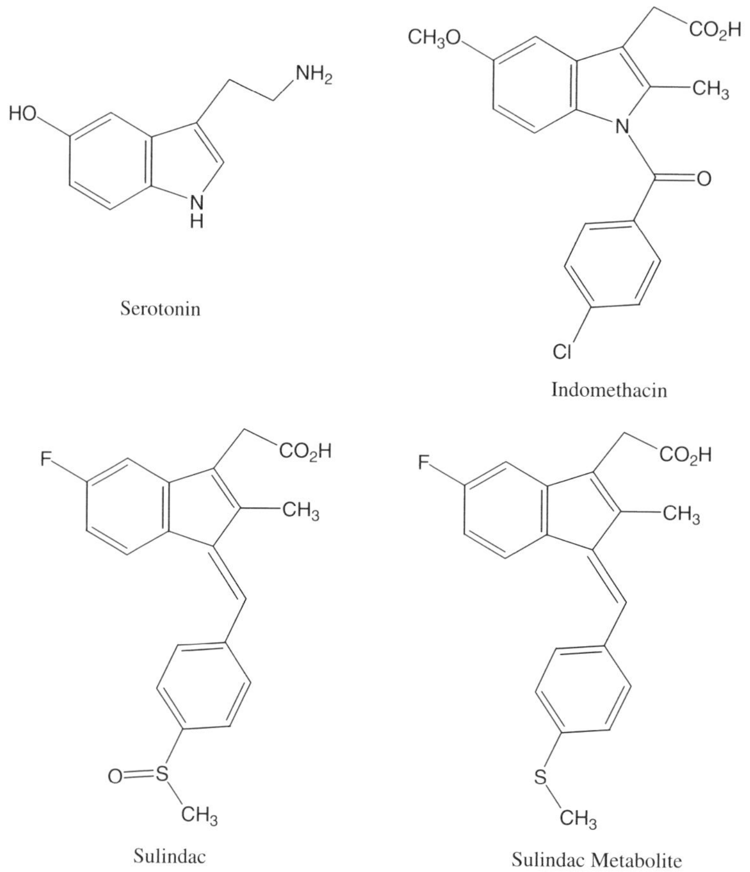

Figure 5   Indomethacin and sulindac.

but dominates the market for nonprescription NSAIDs. Adding a simple methyl group to the acetic acid side chain in ibufenac formally converts that side chain to a propionic acid. The resulting drug, ibuprofen, was tested in the clinic in the United States by Upjohn under license to Boots. The drug was approved by the FDA in 1974. Launched on the market under the name Mortrin®, the compound became an instant runaway success. The Upjohn Company had to scour the world for a source of raw materials for synthesizing the active drug substance. The safety and freedom from side effects, GI ulcers excepted, led to relatively early approval of ibuprofen for over-the-counter use. The drug is currently available under any number of labels because the patent has long since expired. Worldwide production of the active drug substance was estimated a few years ago to exceed 8000 tons.[18] The unexpected commercial success of this drug led to many other laboratories undertaking their own research programs on propionic-acid-based NSAIDs. It is beyond the scope of this book to discuss in detail all the compounds of this class that have been assigned nonproprietary names; the

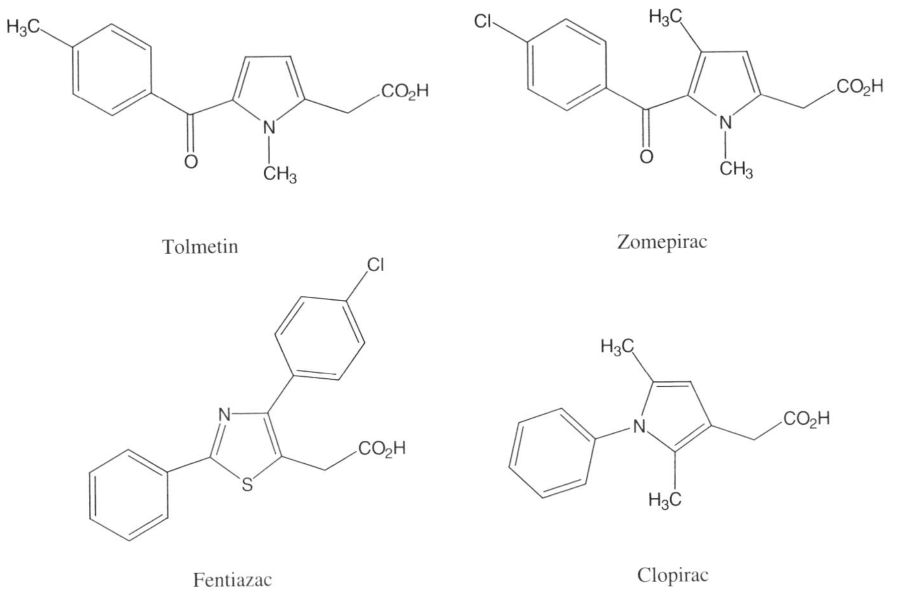
H3C
CO2H
N
O
CH3
Tolmetin
Cl
H3C
CO2H
N
O
CH3
Zomepirac
Cl
N
CO2H
S
Fentiazac
H3C
N
CO2H
H3C
Clopirac

Figure 6 Acetic Acid NSAIDs based on five-membered heterocycles.

Merck index, for example, lists no fewer than 22 anti-inflammatory agents in this propionic acid category. Several representative examples are shown in Figure 7. Naproxen[19] may well rank next to ibuprofen itself in tonnage production. This drug, under the name Aleve®, is widely used as a chronic treatment for arthritis because, unlike relatively short-acting ibuprofen, it need be taken only twice per day. Flurbiprofen,[20] which has a fluorine atom on the ring and in which an additional benzene replaces the isobutyl group, was also first synthesized in the Boots laboratories and is a more potent version of the original drug. Ketoprofen[21] represents one of the many follow-on propionic acids. These drugs are commonly referred to generically as "profens." This appellation goes back to the protocol used by the U.S. Adopted Name Council (USAN) for assigning so-called generic names. Starting several decades ago, this board required that names of all drugs from a specific pharmacologic and structural class use a fixed suffix that would quickly inform practitioners as to the class. Names ending in profens thus denote propionic acid NSAIDs. Naproxen and for that matter ibuprofen predated that convention. By the same token the "-ac"suffix denotes the acetic acid NSAIDs discussed earlier.

The addition of the methyl group to the side chain in the profens means that these molecules, unlike the acetic acid NSAIDs, now exist as a pair of mirror image forms. As is the case with most compounds that act on biological sites, one of those forms accounts for virtually all of the activity. An unusual feature of the profen class is that the less active form is actually converted in the body to the isomer responsible for most of the activity.[22]

Although the NSAIDs discussed to this point were all quite well tolerated, long-term use was clearly associated with stomach irritation and an increased

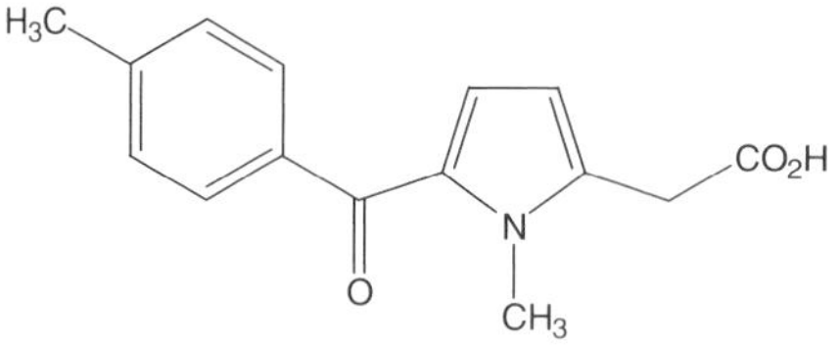

Tolmetin 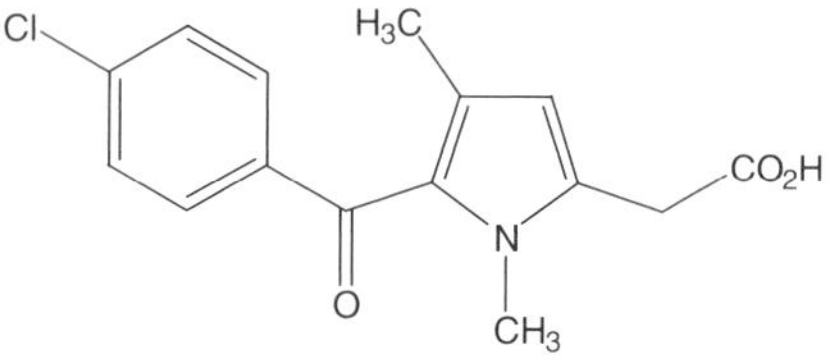   Zomepirac

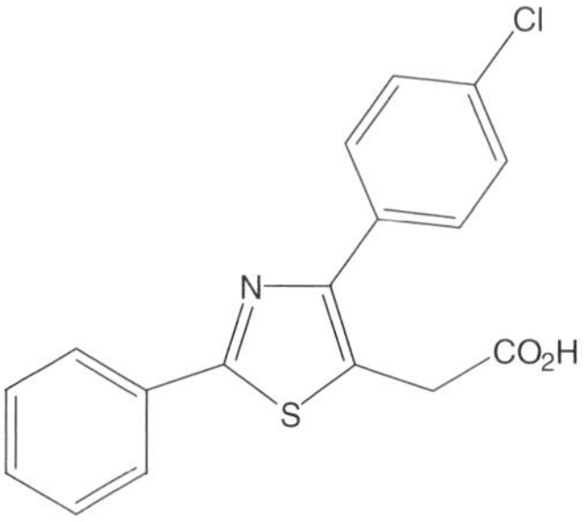

Fentiazac 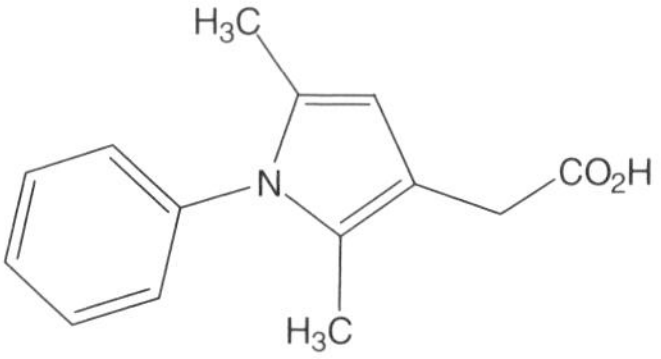   Clopirac

Figure 7   Selected profens.

incidence of stomach ulcers. All theses drugs of course owed their activity to the presence in the structure of a relatively strongly acidic carboxylic acid function, whether attached directly to a benzene ring or in the form of an acetic or propionic. This led to the not unreasonable proposition that this function was in some way related to GI irritation. Chemists at Pfizer thus launched a program to develop NSAIDs that had a somewhat less acidic function.[23] The venerable drug phenyl-butazone, as noted earlier, relies on a proton on carbon linked to two carbonyl atoms for its acidic function. The program resulted in the development of a structu-rally quite distinct class of drugs. The acidic proton in first drug in this category, piroxicam,[23] is also located on a carbon atom atom flanked by two carbonyl groups, in this case a ketone (C=O) and an amide (CONH). This compound, like its predecessor, also undergoes ketone to enol conversion. Piroxicam is usually depicted as one of its enol isomers. This compound was also less acidic than the preceding classes of NSAIDs. Many medicinal chemists are aware of the fact that some pieces of the chemical structures of compounds that have biological activity can be replaced by others that have much the same shape and electron density. These so-called isosteric relationships form part of the lore of the trade. Benzene rings can, for example, be replaced by thiophene. The sulfur atom in this five-membered ring is roughly the same size as the two carbons it replaces; the electron density of this ring is also quite similar to a benzene ring. This relations holds up in the case of tenoxicam,[24] an NSAID that has much the same activity as piroxicam (Fig. 8). The pyiridine ring on the amide function can be replaced by five-membered rings in which two of the carbons in pyridine are replace by either oxygen or sulfur.

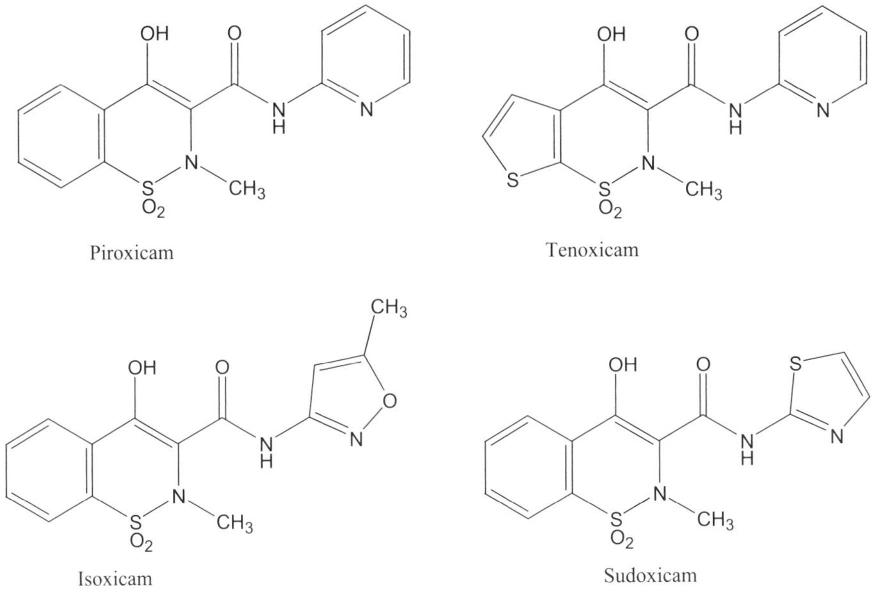

Figure 8 Piroxicam, isoxicam, sudoxicam and tenoxicam.

The resulting compounds isoxicam and sudoxicam[23] are effective NSAIDs in their own right.

   Research that led eventually to elucidation of the mechanism by which these drugs relieved pain and inflammation actually started well before this period, in the early 1960s. This work also eventually resulted in the development of a new class of NSAIDs quite devoid of an acidic proton. The key to some of this work traced back even earlier. Kurzok and later vonEuler had in the 1930s reported the presence of a substance in seminal fluid that contracted the uterus.[25,26] von Euler named this substance prostaglandin for its putative source in the prostate. Vesiculo-glandin or seminaloglandin might have been more apt, as its presence in seminal fluid was later shown to originate in the seminal vesicles. This observation lay fallow for the next three decades because the technology for separating and studying the structure of trace constituents in biological fluids and tissues was not yet up to the task at hand. The compound was finally isolated and its structure established in the early 1960s by Bergstrom and his colleagues at the Karolinska Institute.[27] They subsequently found that prostaglandin was in fact but one of a number of structurally closely related compounds that were widely distributed in the body. Both the compound's potency and activity on isolated muscle experiments promised that this might be a new class of hormones. This suggestion was eagerly seized upon by a company with which Bergstrom had a close working relationship. The Upjohn Company, which was one of the pioneers in developing drugs based on the steroid nucleus, launched a program on prostaglandins that was to persist for decades. Work from that laboratory as well as in many in universities and research institutes elucidated the path by which prostaglandins are formed as well as their functions. Vane and his colleagues at the Wellcome Foundation found that most prostaglandins are in fact agents of injury rather than a new class of hormones.[29] Some of the prostaglandins and their precursors are the very direct cause of inflammation and the attendant pain.

The process by which these compounds are produced is usually referred to as the arachidonic acid cascade. This is shown in outline form in Figure 9 as this set of reactions bears directly on inflammation and the mechanism of NSAIDs. The process starts with the release of the highly unsaturated arachidonic acid from tissue stores catalysed by the enzyme phospholipase A. Another enzyme, cyclo-oxygenase, then adds the elements of oxygen (O_2) to arachidonic acid, leading to the formation of a five-membered ring. This intermediate then undergoes spontaneous transformation to thromboxanes. This last compound is a potent cause of tissue inflammation. Several further steps then give products such as prostaglandin E_1. An entirely different pathway causes the synthesis of a class of open-chain derivatives called eicosanoids. Those compounds are involved in allergic reactions.

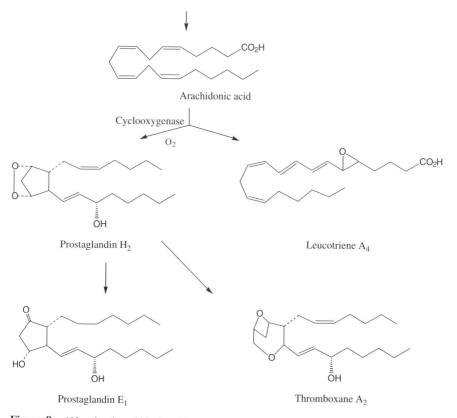

Figure 9   Abbreviated arachidonic acid cascade.

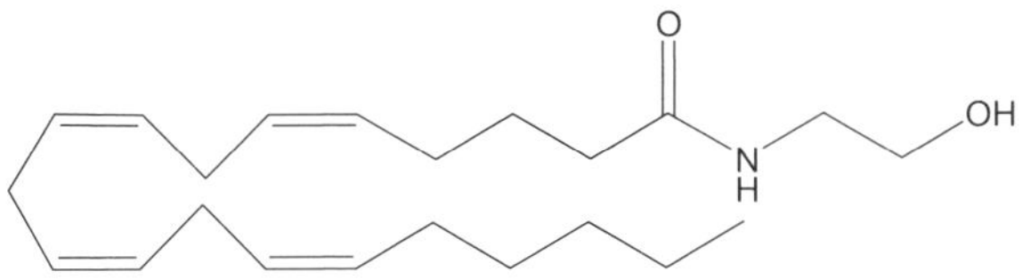

Figure 10   Anandamide.

It might be mentioned in passing that arachidonic acid has recently been found to play a role in the central nervous system as well. The very potent activity of the active ingredient in marijuana, tetrahydrocanabinol, led to a long search for a receptor for this compound. Once that had been identified, the search turned to efforts to find the endogenous compound that interacted with that receptor. The recently identified compound anandamide (Fig. 10) is in fact simply the ethanolamine amide of arachidonic acid.[29] This has spurred efforts to identify both compounds with structures that activate that receptor as well as antagonists, that would antagonize the effects of marijuana.

The detailed picture of how the inflammation-causing products of the arachidonic cascade were generated led in turn to a search to find means by which that pathway might be interrupted. This work led to the perhaps surprising finding that the mother of all NSAIDs, aspirin, in fact quite effectively inhibited the action of the cyclooxygenase (COX) enzyme, sometimes called prostaglandin synthetase. Interference with the action of the enzyme in effect reduced levels of thromboxanes and other pro-inflamatory prostaglandins.[30] This finding was not confined to aspirin; all other known NSAIDS, it was found, in fact owed their activity to inhibition of the COX enzyme. The use of the singular for the latter substance reflects the state of knowledge at that time.

More detailed research later showed that the ubiquitous products of the arachidonic acid cascade play beneficial as well as injurious roles. One of the latter functions helped explain the most prominent side effect of NSAIDs as a class, that is, their propensity to lead to the development of gastric and duodenal ulcers. This starts with the fact that the stomach is amply supplied with highly acidic juices that contain enzymes such as peptidases and proteases, a combination that is required for digesting proteins. The specialized layer of cells and mucus that line the interior of that organ in effect protects its own proteinaceous structure from those digestive juices. Prostaglandins, it was found, play a very direct role in maintaining that protective layer; this is especially important in secretion and maintenance of the mucus layer. Inhibition of the COX enzyme decreases levels of prostaglandins and thus the very necessary protective layer. The very mechanism by which NSAIDs act at the same time accounts for their salient side effect. The stomach irritating and ulcerogenic activity of this class thus turns out to be a simple extension of their mode of action.[31] Locally administered prostaglandins could at least in theory overcome that side effect. The compounds in the natural series are, however, too quickly destroyed in the stomach to be of any use. Chemists at G. D. Searle were able to synthesize a congener that survived

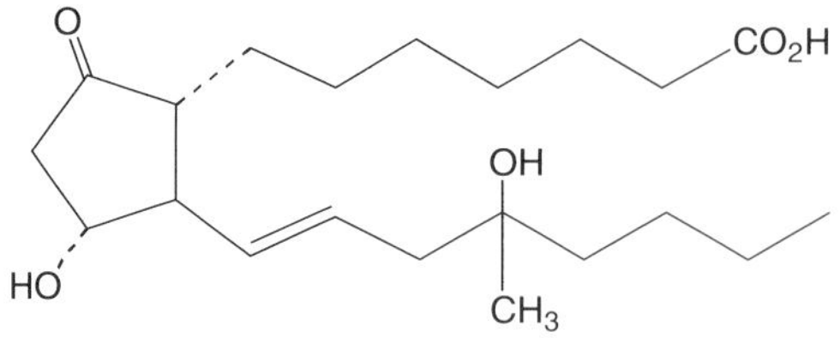

Figure 11   Misoprostol.

that hostile environment. This drug misoprostol (Fig. 11) is indicated for reducing the propensity for NSAID-induced ulcers in patients in high-risk populations such as the elderly on high-dose chronic anti-inflammatory regimes or individuals with a previous history of ulcers. The drug is also used off-label as an adjunct to the arbotifacient mifepristone (RU-486) that is discussed in the chapter on steroids. Misoprostol is virtually the only prostaglandin-derived drug to have achieved much of a market in spite of the enormous expenditure of time and money devoted to the field in both the private and public sectors.

Research of NSAIDs had pretty much run its course by the 1990s. The close mechanistic connection between activity and the side effect seemed to preclude the likelihood of finding a drug that would divorce those properties. Earlier discussions in this volume have noted that more detailed experiments on enzymes have time and again led to the finding that they exist in subcategories. Such findings also sometimes led to the discovery of new classes of drugs. This has recently also been the case with NSAIDs. Research on the DNA that codes for the synthesis of COX led to the discovery that this enzyme too existed in two forms.[32] These were designated COX-1 and COX-2, forsaking in this case the use of Greek letters. It was found within a short time that COX-1 is the main enzyme responsible for maintaining the gastric mucosa, and both it and COX-2 were involved in prostaglandin production at most other sites.[33] A drug that was selective for COX-2 would at least in theory have anti-inflammatory activity without the attendant effect on the gastric mucosa.

It might be noted in passing that some quite recent research has led to the identification of yet another variation of this enzyme. That subcategory, COX-3, is said to act at only the analgesic and fever-causing sites. This would account for the activity of drugs such as acetaminophen, which are devoid of anti-inflammatory activity.[34]

The structural lead for the class of drugs that are selective for COX-2 actually came from a quite unexpected source that traced back almost half a century.[35] Very few practical sources of steroids were available when the sex steroids were discovered in the 1930s. As noted in the chapter on steroids, this led to the development of the nonsteroid estrogens. The nonsteroid estrogen antagonists comprised a further extension of that research. The announcement by a group at Richardson–Merrell that one of those compounds, MRL-41 (Fig. 12), showed good antifertility activity in rats led a number of competitors to launch their own programs based on that structural lead. Although it was known by then that a steroid oral contraceptive was in the late stages of development, it was thought that a nonsteroid might

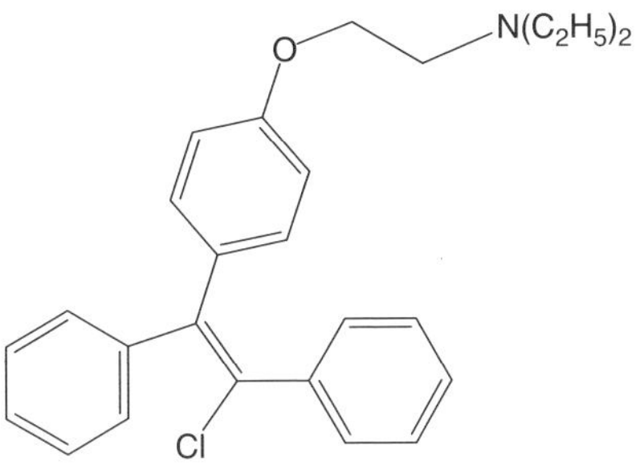

Figure 12   Nonsteroid estrogen antagonists.

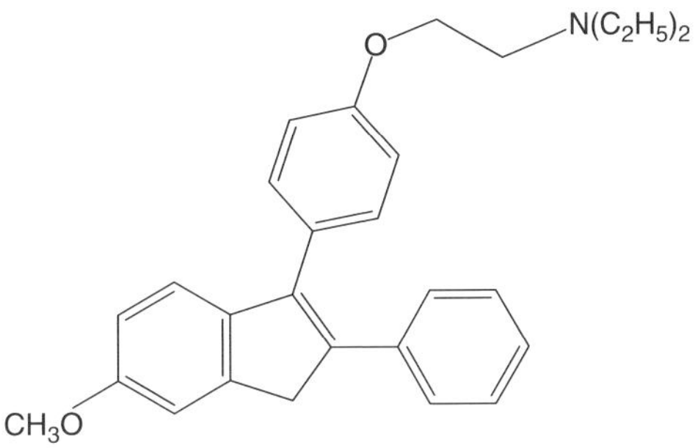

have cost or other advantages. A very small group at Upjohn decided to investigate the effect of closing up part of the MRL-41 molecule as a ring. This aim of this maneuver was to produce a compound, called a rigid analog, that would better fit the putative unknown receptor. The ultimate goal of the work was a series of compounds with a pair of fused six-membered rings. These, it was felt, would resemble the AB rings in a steroid. The initial series, however, comprised a five-membered ring fused to benzene,[36] an indene, because these were accessible in a few steps and starting materials were readily available. The later six-membered analogs and their history are covered in the chapter on steroids. The chemist who prepared those compounds was gratified to find that the very first compound in the series showed activity in the screen. This encouraged expansion of the lead by synthesizing additional analogs. The series of compounds that resulted from the program culminated in the analogue U-11,555, a compound that was much more potent in the screen than MRL-41. This was then slated for clinical trial as an oral contraceptive. After it cleared toxicology assays, the compound was tested for tolerability in Phase I trials in human volunteers. Reports came back very quickly that the agent was quite phototoxic. Subjects who had taken the compound developed extreme sensitivity to sunlight.

Another group at Upjohn was heavily involved in indole chemistry at about the same time, arguably motivated by the presence of that group in many biologically active molecules such as serotonin. This same rationale, as noted earlier, led to the discovery of indomethacin. The discovery of the indene antiestrogen U-11,555 seems not to have escaped their notice. Accordingly they prepared a series of indoles with benzene rings attached to the same positions in the five-membered ring as in the indene contraceptive. Upjohn at the time maintained a high-throughput screen for anti-inflammatory compounds, largely as a result of their still ongoing large program on steroids. The screen also accepted any compounds from other structural classes that were available in sufficient quantity. One of the indoles from the series with the two extra rings unexpectedly showed good activity in the screen as well as a battery of follow-up tests.[37,38] The activity was difficult to explain at that time as the compound lacked any trace of the acid proton present in other known NSAIDs.

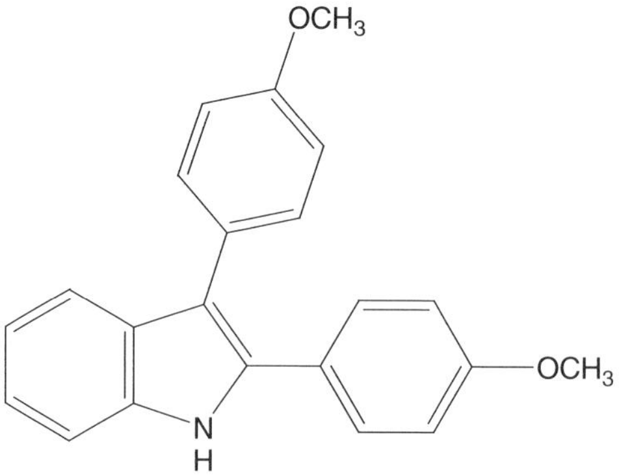

Indoxole

Figure 13   Indoxole.

This indole was taken to the clinic under the nonproprietary name indoxole (Fig. 13). This agent also failed in Phase 1 testing; very much like its indene precursor the compound also caused severe photosensitivity.[39]

There was understandably great reluctance to walk away from this lead, as indoxole had shown outstanding activity in various animal models. The fact that the structure of the compound was not related to any of the then known anti-inflammatory drugs was added motivation to try to salvage the lead. The company thus launched a major program to ring modifications on the molecule to attempt to produce nonphototoxic analogs. This initiative also required finding an animal model for the side effect, as the commonly furred laboratory species did not manifest phototoxicity. The program gave very disappointing results. The majority of the modifications on the indoxole structure led to loss of anti-inflammatory activity. Those few indoles that retained activity were still phototoxic. This presented the inescapable conclusion that the toxicity in both indoxole and U-11,555 was due to a shared feature: the presence of the two benzene rings on the double bond in the benzene-fused five-membered ring.

There matters would have rested had it not been for a modest freelancing program intended to mimic the structure of indoxole with compounds in which the fused ring was replaced by heterocycles, that is, fragments that included atoms other than carbon. Modest anti-inflammatory action was interestingly retained when the fused benzene ring was replaced by a five-membered sulfur and nitrogen-containing fragment (imidazothiazole) or one in which oxygen replaced the sulfur in the latter (imidazooxazole) (Fig. 14).[40] These replacements are generally in line with the isosteric equivalents noted earlier. Perhaps most surprising was the finding that the fused ring could be dispensed with entirely. The compound in which the two benzene rings were attached to adjacent carbon atoms on a sulfur- and nitrogen-containing five-membered thiazole ring[41] also showed modest anti-inflammatory activity. The presence of methoxy groups (CH_3O) on the benzene seemed to be a stringent requirement for activity in this small series. The level of biological activity in these series was judged too low to warrant follow-up at that time.

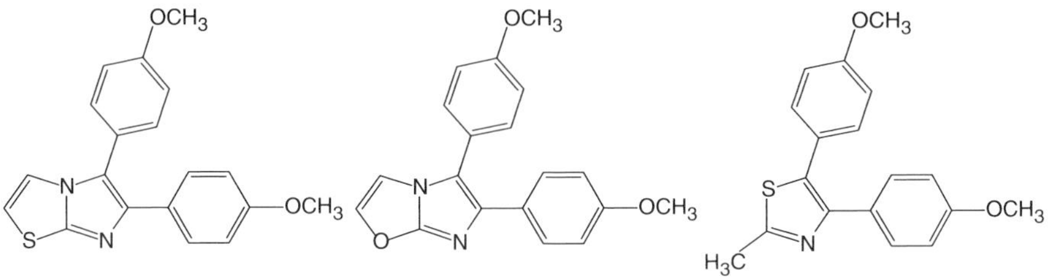

Figure 14   Fused and single ring heterocyclic anti-inflammatories.

Blood platelets are one of the many circulating white cells. Their specific function involves initiation of the process that leads to the formation of blood clots. Any break in the wall of a blood vessel will cause platelets to aggregate at that site. This sets off the long cascade of events that culminates in the formation of a clot. When platelets aggregate in the absence of injury, the circulating aggregates or even clots can lead to blockage of vital blood vessels, an event more familiarly known as a stroke. Upjohn, in common with many other companies, set up a program in the middle 1970s to screen compounds from their chemists to identify compounds that inhibited platelet aggregation. It is relevant to note in passing that one of the many activities of NSAIDs, and aspirin in particular, is the ability to inhibit platelet aggregation. This accounts for both the recommendation for half an aspirin per day to fend off strokes and the prohibition for consuming NSAIDs within a week of surgery. The new screen was, however, intended to uncover new classes of compounds rather than those known to have that activity as an ancillary property. As is common with most high-throughput screens, test candidates included a good proportion of compounds from the collection of abandoned substances. The finding that the thiazole that had come out of the indoxole program showed reasonable activity in this new screen resulted in the synthesis of new analogs. This project culminated in a compound in which the protons on the methyl group on the original lead were replaced by fluorine.[42] This agent showed sufficient activity to be groomed for further development. It was taken to the clinic as a potential platelet anti-aggregating agent under the generic name itazigrel (Fig. 15).[43] Although apparently not tested for its effect on inflammation, one report did allude to the drug's effect on cyclooxygenase production.[44] DuP-697 is one of the first compounds in this general structural class targeted at inflammation.[45] The structure retains the same general basic ring system as the original thiazole, that is, two benzene rings on a five-membered heterocycle. The central ring is now a thiophene, in which sulfur is the only atom other than carbon. In addition, one of the methoxy groups has been replaced by a fluorine group and the other by a sulfonylmethyl function (SO_2CH_3); bromine replaces the methyl (or trifluormethyl) group. The compound was later found to be a specific COX-2 inhibitor, although the work preceded the identification of the COX subtypes. DuP-697 was abandoned because of its poor pharmacokinetic properties.[46]

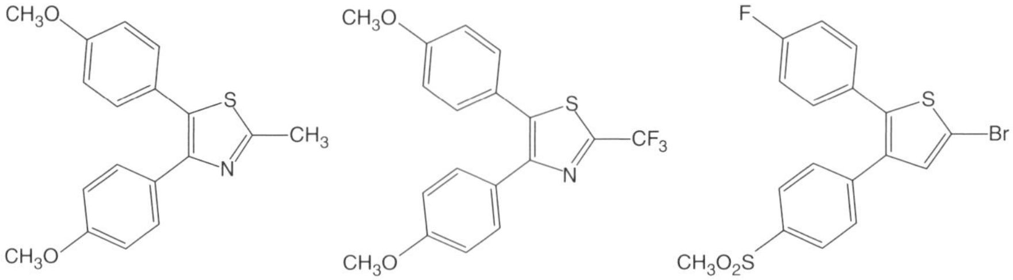

| Thiazole | Itazigrel | DuP-697 |

Figure 15 Thiazole COX inhibitors.

The next development in what came to be a very active field of research came from the Searle laboratories. The structure of the compound celecoxib follows the same very general overall pattern, with some significant changes (Fig. 16). The central ring is a pyrazole that includes a pair of adjacent nitrogen atoms.[47] The presence of the sulfonamide function (SO_2NH_2) is quite suggestive. The hydrogen atoms on this group have about the same acidity as a phenol. It is not unlikely that at least one of the methyl ether groups in itazagrel is cleaved in the body to a phenol proper (OH). Celecoxib proved to be a selective COX-2 inhibitor. Once it passed clinical trials and was approved by the FDA, it went on to become a very successful drug. This agent, Celebrex®, promised an NSAID with reduced risk of ulcers on long-term use. The same laboratory successfully developed the structurally related COX-2 inhibitor valdecoxib, based on a heterocycle that includes nitrogen and oxygen called an isoxazole. This too includes the somewhat acidic sulfonamide function.[48] That compound is now approved and marketed as Bextra®. Parecoxib is a derivate of valdecoxib in which the sulfonamide group has been modified to increase the water solubility of the drug. This agent is intended for intravenous use for treatment of postoperative pain.

A rather different approach was taken by scientist at Merck. The central five-membered ring of rofecoxib is still formally a heterocycle.[49] This fragment can also be viewed as the internal ester, that is the lactone of an unsaturated hydroxyacid. The substituent on the benzene ring in rofecoxib is a sulfonyl methyl group as in DuP-697 rather than the sulfonamide in celecoxib. This drug, marketed as Vioxx®, was widely prescribed for treatment of osteoarthritis until the fall of 2004. At that time Merck withdrew the compound from the market because of reports of excessive deaths among individuals with heart disease who were taking Vioxx. The ensuing press coverage cast a pall on COX-2 NSAIDs as a class. This also brought to the fore some earlier reports that noted that the difference of the rate of ulcers among users of these drugs and those on the classic NSAIDs was at best very small.

Two other recently reported compounds described as COX-2 inhibitors have structures that differ significantly from the pattern of their forerunners.

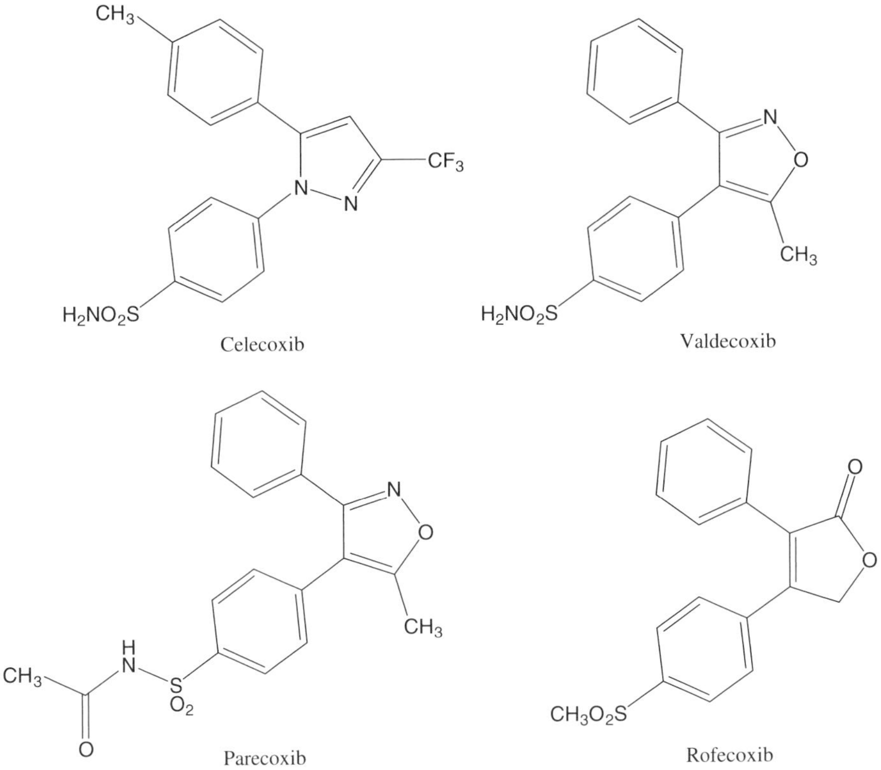

Figure 16 Five-membered COX-2 inhibitors.

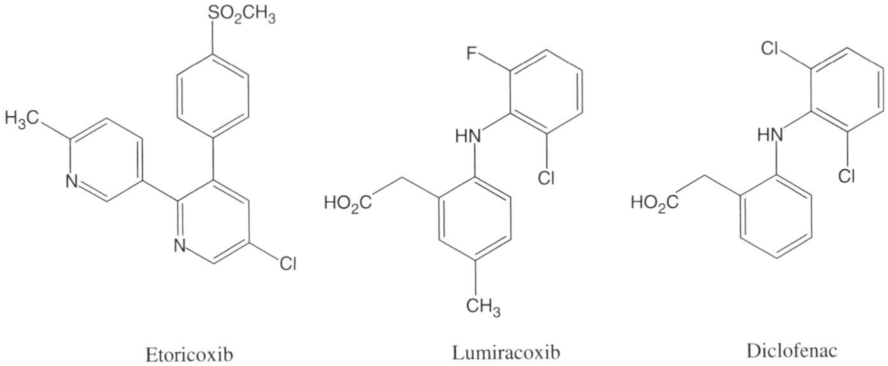

Figure 17 Miscellaneous COX-2 inhibitors.

A six-membered, nitrogen-containing pyridine ring replaces the customary five-membered heterocycle in the COX-2 inhibitor etoricoxib (Fig. 17).[50] The compound does retain the methyl sulfonyl group of its predecessor rofecoxib. The structure of lumiracoxib[51] represents the most drastic departure from what had been the traditional pattern assumed to be required for drugs that preferentially targeted COX-2. The presence of an acetic acid side chain in this molecule in the compound may easily be construed as having come back full circle to the older "-ac" suffixed NSAIDs. Comparison of the structure of this compound with that of the venerable NSAID diclofenac reveals only minor differences: replacement of a chlorine by fluorine and the presence of an additional methyl group in one benzene ring.

REFERENCES

1. C. E. Buss, *Deut. Arch. Klin. Med.*, 15, 457 (1875).
2. For a review of work up to 1982 see J. G. Lombardino, in J. G. Lombardino, Ed., *Nonsteroidal Antiinflammatory Drugs*, Wiley, NY, 1985, pp. 251–431.
3. J. H. Wilinson and I. L. Finar, *J. Chem. Soc.*, 32 (1948).
4. C. V. Winder, J. Wax, L. Scotti, R. A. Scherrer, E. M. Jones and F. W. Short, *J. Pharmacol. Exp. Ther.*, 138, 405 (1962).
5. M. H. Scherlock and N. Sperber, U.S. patent 3,337,570 (1967).
6. H. Stenzl, U.S. patent 2,562,830 (1951).
7. R. Pfister and F. Hafliger, *Helv. Chim. Acta.*, 40, 395 (1957).
8. A. Sallmann and R. Pfister, U.S. patent 3,558,690 (1971).
9. D. C. Atkinson, K. E. Godfrey, B. J. Jordan, E. C. Leach, B. Meek, J. D. Nichols and J. F. Saville, *J. Pharm. Pharmacol.*, 26, 357 (1974).
10. T. Y. Shen, R. L. Ellis, T. B. Windholz, A. R. Matzuk, A. Rosegay, S. Lucas, B. E. Witzel, C. H. Stammer, A. N. Wilson, F. W. Holly, J. D. Willet, L. H. Sarett, W. J. Holtz, E. A. Rislay, G. W. Nuss and C. A. Winter, *J. Am. Chem. Soc.*, 85, 488 (1963).
11. T. Y. Shen, R. B. Greenwald, H. Jones, B. O. Linn and B. E. Witzel, U.S. patent 3,654,349 (1972).
12. D. E. Duggan, K. F. Hooke, E. A. Risley, T. Y. Shen and C. G. Arman, *J. Pharmacol. Exp. Ther.*, 201, 8 (1977).
13. J. R. Carson, D. N. McKistry and S. Wong, *J. Med. Chem.*, 14, 664 (1971).
14. J. R. Carson and S. Wong, *J. Med. Chem.*, 16, 172 (1973).
15. G. Lambelin, J. Roba, C. Gillet and N. P. Buu-Hoi, German patent 2,261,965 (1973).
16. K. Brown, D. P. Cater, J. F. Cavalla, D. G. Green, R. A. Newberry and A. B. Wilson, *J. Med. Chem.*, 17, 1177 (1974).
17. J. S. Nicholson, in J. S. Bindra and D. Lednicer, Eds, *Chronicles of Drug Discovery*, Vol. 1, Wiley, NY, 1982, pp. 149–172.
18. R. A. Sheldon, *Pure Appl. Chem.*, 72, 1232 (2000).
19. I. T. Harrison, B. Lewis, P. Nelson, W. Rooks, A. Roszowski, A. Tomalonis and J. H. Fried, *J. Med. Chem.*, 13, 203 (1970).
20. S. S. Adams, J. S. Bernard, J. S. Nicholson and A. B. Blancafort, U.S. patent 3,755,427 (1973).
21. D. Furge, N. M. Messer and C. Moutonnier, U.S. patent 3,641,127 (1972).
22. J. M. Mayer, M. Roy-De Vos, C. Audergon, B. Testa and J. C. Etter, *Int. J. Tissue React.*, 16, 59 (1994).
23. E. H. Wiseman and J. G. Lombardino, in J. S. Bindra and D. Lednicer, Eds, *Chronicles of Drug Discovery*, Vol. 1, Wiley, NY, 1982, pp. 173–200.

24. D. Bradshaw, C. H. Cashin, A. J. Kennedy and N. A. Roberts, *Inflamm. Res.*, 15, 569 (1984).
25. R. Kurzok and C. C. Lieb, *Proc. Soc. Exp. Biol. Med.*, 28, 268 (1930).
26. U. S. vonEuler, *J. Physiol. (London)*, 4, 213 (1937).
27. S. Abrahamson, S. Bergstrom and B. Samuelson, *Proc. Chem. Soc.*, 352 (1962).
28. R. Vane, Nobel Lecture, *Br. J. Pharmacol.*, 79, 821 (1983).
29. W. A. Devane, L. Hanus, A. Breuer, R. G. Pertwee, L. A. Stevenson, G. Griffin, D. Gibson, A. Mandelbaum, A. Etinger and R. Mechoulam, *Science*, 258, 1882 (1992).
30. G. A. Higgs, F. A. Harvey, S. H. Ferreira and J. R. Vane, *Adv. Prostaglandin Thromboxane Res.*, 1, 105 (1976).
31. B. J. Whittle and J. R. Vane, *Arch. Toxicol.*, Suppl. 7, 315 (1984).
32. T. Hla and K. Neilson, *Proc. Natl. Acad. Sci. U.S.A.*, 89, 7384 (1992).
33. I. A. Tavares and A. Bennett, *Int. J. Tissue React.*, 15, 49 (1993).
34. B. Kis, J. A. Snipes and D. W. Busija, *J. Pharmacol. Exp. Ther.*, 315, 1 (2005).
35. Much of the material that follows appeared in D. Lednicer, *Curr. Med. Chem.*, 9, 1457 (2002); used by kind permission from Bentham Science Publishers.
36. D. Lednicer, J. C. Babcock, P. E. Marlatt, S. C. Lyster and G. W. Duncan, *J. Med. Chem.*, 8, 52 (1965).
37. J. Szmuszkovicz, E. M. Glenn, R. V. Heinzelman, J. B. Hester and G. A. Youngdale, *J. Med. Chem.*, 9, 527 (1966).
38. E. M. Glenn, B. J. Bowman, W. Kooyers, W. Koslowske and M. L. Myers, *J. Pharmacol. Exp. Ther.*, 155, 157 (1967).
39. A. D. Steele and D. J. McCarty, *Ann. Rheum. Dis.*, 26, 39 (1967).
40. D. Lednicer, U.S. patent 3,455,924 (1968).
41. D. Lednicer, U.S. patent 3,458,526 (1969).
42. R. H. Rynbrandt, E. E. Nishizawa, D. P. Balgoyan, A. R. Mendoza and K. A. Annis, *J. Med. Chem.*, 24, 1507 (1981).
43. D. M. Demke, J. R. Luderer, L. K. Wakefield, A. R. Euler, R. D. Brouwer and C. M. Metzler, *J. Clin. Pharmacol.*, 27, 916 (1987).
44. P. H. Hsyu, M. D. Koets and J. R. Luderer, *J. Pharm. Sci.*, 83, 1747 (1994).
45. K. R. Gans, W. Galbraith, R. J. Roman, S. B. Haber, J. S. Haber, J. S. Kerr, W. K. Schmidt, C. Smith, W. E. Hewes and N. R. Ackerma, *J. Pharm. Exp. Ther.*, 254, 180 (1990).
46. D. J. Pinto, private communication.
47. For a review see J. J. Talley, in F. King and A. Oxford, Eds, *Progress in Medicinal Chemistry, 36*, Elsevier, Amsterdam, 1999, pp. 201–234.
48. J. J. Talley, S. R. Bertershaw, D. L. Brown, J. S. Carter, M. J. Granetto, M. S. Kellog, C. M. Kobolett, J. Yuan, Y. Y. Zhang and K. Seifert, *J. Med. Chem.*, 43, 1661 (2000).
49. P. Prasit, Z. Wang, C. Brideau, C. C. Chan, S. Charleson, W. Cromlish, D. Ethier, J. F. Evans, A. W. Ford-Hutchinson, J. Y. Gauthier, R. Gordon, J. Guay, M. Gresser, S. Kargman, B. Kennedy, Y. Leblanc, S. Leger, J. Mancini, G. P. O'Neill, M. Ouellet, M. D. Percival, H. Perrier, D. Riendeau, I. W. Rodger, P. Tagari, M. Therien, P. Vickers, E. Wong, L.-J. Xu, R. N. Young and R. Zamboni, *Bioorg. Med. Chem. Lett.*, 9, 1773 (1999).
50. R. W. Friesen, C. Brideau, C. C. Chan, S. Charleson, D. Deschenes, D. Dube, D. Ethier, R. Fortin, Y. Gauthier, Y. Girard, R. Gordon, G. M. Greig, D. Riendeau, C. Savoie, Z. Wang, E. Wong, D. Visco, J. Xu and R. N. Young, *Bioorg. Med. Chem. Lett.*, 8, 2777 (1998).
51. C. Ding and G. Jones, *Drugs*, 5, 1168–1172 (2002).

Chapter 7

Steroids

This discussion of the development and discovery of drugs has to this point more or less neatly fallen into therapeutic categories. Each of the preceding chapters traced the discovery of new chemical compounds for treating broad disease categories. This involved not only the different pharmacological mechanisms used for approaching treatments for the disorder, but also the diverse classes of chemical structures of drugs that acted on those mechanisms. There remains, however, a broad class of drugs that is better treated in terms of the common chemical structure. The four fused-ring steroid nucleus provides the framework for the major classes of hormones that regulate many life processes. A sizeable number of drugs that have no other unifying theme are either based directly on the steroid nucleus or trace their lineage to that class of endogenous compounds.

Much of the early history of steroids relates to the two compounds abundantly available from natural sources, cholesterol and the cholic acid. These have been known since antiquity, the former as the main constituent of gall stones and the latter as the principal ingredient of bile. Cholesterol itself occurs in surprisingly high concentrations in nervous tissue and the brain. Spinal cord of cattle at one time constituted the main source of this compound. It was obtained in pure crystalline form early in the nineteenth century, although the correct empirical formula, $C_{27}H_{46}O$, was not established until 1888. Both this steroid and cholic acid played a key role in determining the correct structural formula for the steroid nucleus.[1] By the early part of the twentieth century, organic chemistry had progressed far enough for chemists to tackle seriously the problem of deciphering the structure of these molecules. The determination of the detailed makeup of these complex molecules nevertheless posed a major challenge.

In the absence of the multitude of instrumental methods available to today's chemists, the work relied virtually entirely on subjecting the compounds to known chemical reactions that would, for example, chop off pieces, then studying the nature of the fragment via further reactions. An alternative involved sequentially opening the rings via oxidation reactions and again divining the nature of the product

New Drug Discovery and Development by Daniel Lednicer
Copyright © 2007 John Wiley & Sons, Inc.

by studying its reactions and their products. In 1919 Adolf Windaus at Gottingen had been able to convert cholesterol to a close derivative of cholic acid, showing that the two compounds had very closely related chemical structures. He and Heinrich Wieland at Munich attacked the problem by carrying out the detailed and exhaustive series of reactions starting with cholanic acid. This culminated in 1928 in a proposal for the structure of the steroid nucleus. This was based heavily on serial degradation reactions and observing the chemical properties of the subsequent products. There was, however, still some question as to the location of one of the methyl groups known to be present in the molecule. An apocryphal story refers to that as the floating methyl group. In the same year they were awarded a joint Nobel Prize for their work on steroids. The timing makes it clear that the prize was given for the body of the work up to that point rather than for the solution of the structure of the steroid nucleus. X-ray crystallography as a tool for the determination of the structure of organic compounds was then still in its infancy. It was not as yet able to provide, as is the case today, the detailed data that pinpoint the position of every atom in a molecule. In 1932 Bernal in the UK had determined by X-ray crystallography that another steroid, ergosterol, consisted of a long flat molecule. The Wieland–Windaus proposal, however, predicted a relatively thick molecule. It was then recalled that some years earlier, Otto Diels had reported that one of the products from heating cholesterol at a very high pressure was the already known planar four-ring hydrocarbon chrysene. This too suggested that steroids consisted of flat molecules. Re-examination of the data from the degradation reaction showed that one step in the long sequence had been misinterpreted. This led to the adoption of the correct formulation of the steroid nucleus, which is indeed a long flat molecule. The four rings and the attached methyl groups and the substituent at the 17 position (see Fig. 1) include six carbon atoms that can exist in one of two mirror image

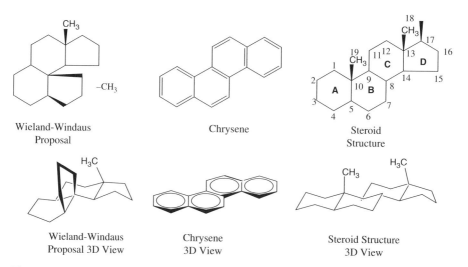

Figure 1   Early and current structures for the steroid nucleus.

forms. This leads to 36 possible isomers; as is the case with most natural products, these occur in natural steroids in but one form, with occasional exceptions at the 17 position. Figure 1 also includes the current numbering system and ring labels, as these will occasionally occur in the accounts that follow. The strange numbering scheme traces back directly to the problems in assigning a structure to the steroid nucleus.

Pharmacology too witnessed major advances in the early part of the twentieth century. Experiments had by then clearly demonstrated that reproductive function in females as controlled by substances secreted by the ovaries; corresponding effects in males responded to secretions from the testes. In 1923, Allen and Doisy found that a cell-free alcoholic extract from the ovaries of rats in estrus would induce estrus when injected into other rats whose ovaries had been removed. This strongly suggested that the factors in the extracts that induced estrus consisted of relatively low-molecular-weight compounds. Further work along this line culminated in 1929 in the isolation of the active principle as a pure crystalline compound by Doisy and his collaborators in the United States and Butenandt in Germany. Structure determination work on this compound, eventually named estrone, followed the same strategy as had been used for cholesterol, that is, sequential degradation reactions. By 1932 it had been shown that the structure of estrone was based on the four-ring steroid nucleus. The same was true for the closely related estrogens also secreted by ovaries, estradiol in which the ketone at 17 is reduced to an alcohol, and estriol, which has yet another hydroxyl group in the five-membered ring (Fig. 2). A major difference between the structure of these compounds and that of cholesterol is that the A ring now consists of a benzene ring. The methyl group that is present at the juncture of rings A and B in cholesterol and other steroids is of necessity missing as well. Isolation and structural determination of the other steroid involved in female reproductive function, progesterone, as well as the male hormone testosterone, followed not much later. These steroid hormones are discussed in more detail further on in this chapter.

It was already recognized in the early 1930s that various "female complaints" were related to low levels of estrogens. The use of these newly identified hormones as supplements to alleviate those conditions was at that time, however, problematic, because the necessary estrogens were not available in a form that could be administered to patients. The very low levels at which compounds occurred in

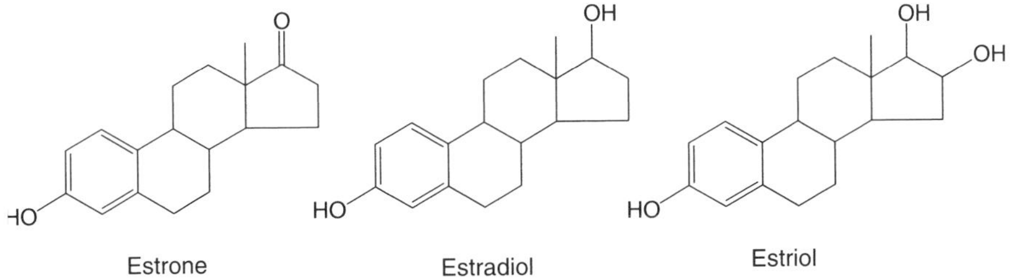

Figure 2   Ovarian estrogens.

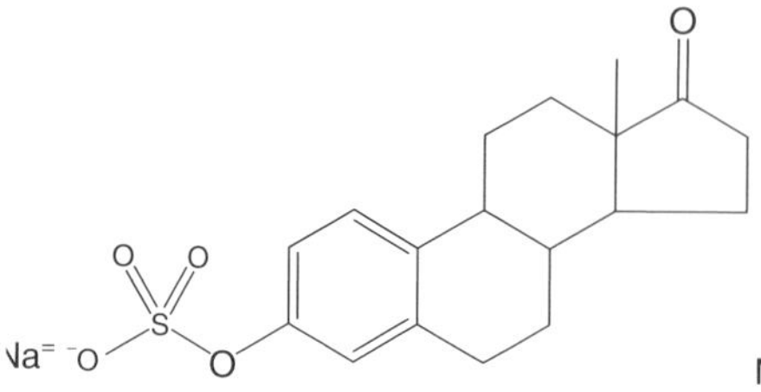

Estrone Sulfate 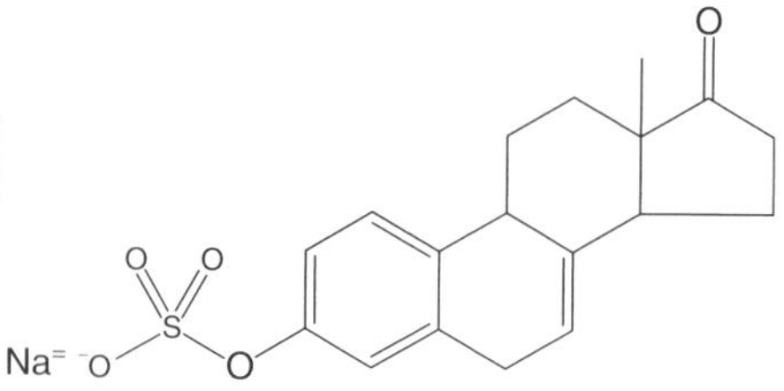   Equilin Sulfate

Figure 3   Conjugated estrogens.

animal tissues made it difficult to use those as a source of the drug. Although steroids occur abundantly in plants, none of those sources included compounds with a benzene A ring. Methods for converting plant-derived steroids to estrogens, as will emerge later, were not developed until well into the 1940s. Three decades were to elapse before the development of chemistry for synthesizing these compounds from coal-tar or petroleum-derived starting materials. These methods, perfected in the 1960s, will also be touched upon later.

One of the early estrogen replacement products was based on the observation that pregnant mare urine was rich in substances that showed estrogenic activity. That urine, in common with that from many other gravid animals, contained high concentrations of estrogen metabolites. An extract of that urine, termed equine conjugated estrogens, was approved for medical use by the FDA in 1942. This extract comprises a mixture of the water-soluble sulfate esters of closely related steroids. When first approved, the drug, named Premarin®, was known to consist largely of the sulfate esters of estrone, and equilin, the name of the latter reflecting its source (Fig. 3). Additional other minor constituents have been identified over the years as analytical chemistry methodology evolved. The role played by the various components of this complex mixture on the biological activity of the product is an open question.[2] Applicants for a license to market nonproprietary versions of a drug must demonstrate that their version has the same active drug substances as the original. The lack of a generic equivalent of Premarin® is probably due at least in part to the complexity of the mixture.

A number of highly simplified versions of the estrone had been synthesized as reference materials in the course of work on the structure determination of the estrogens. Some of those compounds, such as a three-ring phenanthrone, surprisingly showed activity in animal tests for estrogens.[3] This finding led to the search for an even more abbreviated version of estrone in the search for a drug that could be readily synthesized with the technology available in the 1940s. One of the compounds to emerge from this program was diethylstilbestrol (DES).[4] This relatively easily prepared compound became the first widely used estrogen. The fact that it was orally active and relatively inexpensive led to its prescription for a host of indications. There was little supporting clinical evidence for some of

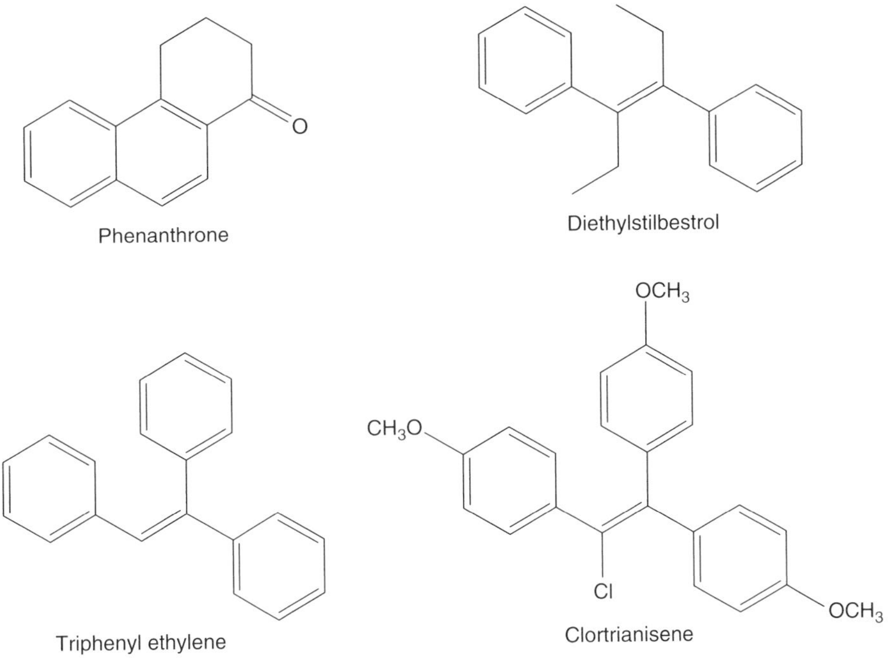

Phenanthrone

Diethylstilbestrol

Triphenyl ethylene

Clortrianisene

Figure 4 Synthetic estrogens.

those uses, such as treatment of threatened pregnancy. The much later finding that ingestion of diethylstilbestrol led to unusual cancers in the daughters of women who had taken the drug while pregnant led to its withdrawal from the market. Subsequent research on these simplified synthetic estrogens had revealed that one of the ethyl groups in DES could be replaced by benzene to give the synthetic estrogen triphenylethylene. A derivative of this last, clortrianisene,[5] also known by its acronym TACE, in which each of the benzene rings has a methoxy group, and chlorine replaces the last hydrogen on the central two-carbon chain, is still on the market.

Circumstances occasionally arise where it is desirable to administer estrogenic agents by injection. The synthetic estrogens (Fig. 4) such as TACE are far too insoluble in water to be given by that route. Chemists at Richardson–Merrell sought to overcome that problem by preparing derivatives of TACE-like molecules that would form salts with strong acids. Those should be more soluble in water than the parent molecules. They thus replaced one of the methyl groups on the methoxyl in a variety of analogs of TACE by a two-carbon chain that carried a basic nitrogen at the end, a so-called basic ether. Salts of those compounds formed by neutralization of basic nitrogen with an acid were, as expected, more soluble in water. They were, however, interested to note that the products were now no longer classical estrogens. Instead, these new derivatives also blocked the effect of concurrently administered estrogens (Fig. 5). Much of the research

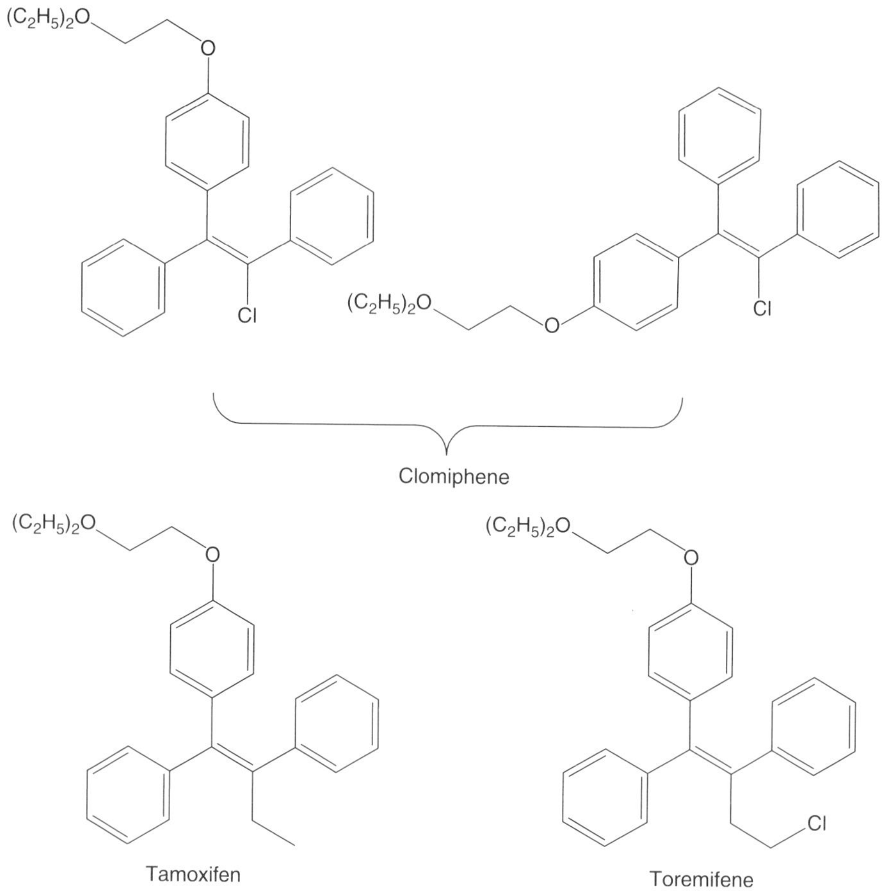

Clomiphene

Tamoxifen

Toremifene

Figure 5 Estrogen antagonists.

on these and related analogs was prompted by the fact that these agents also acted as oral contraceptives in laboratory rodents. This property turned out to be due to a peculiarity of the reproductive cycle in those species. One of the series, clomiphene, also known as MRL-41,[6] was placed on the market for treatment of infertility due to ovulatory problems. This drug gained some fame by its associating with multiple births. This compound can exist as two isomers, one in which chlorine and the basic ether-bearing benzene are on the same side of the two-carbon double-bonded chain and that in which they are on opposite sides. The commercial product comprises a fixed mixture of those two isomers.

It was well known by the 1960s that the growth of many breast cancers was stimulated by endogenous estrogen. Treatment of the disease not infrequently involved removal of a woman's ovaries in order to shut off that source of the hormone. Scientists at ICI in the UK followed up on the discovery of the estrogen antagonists in order to achieve the same goal pharmacologically. Their research led to the synthesis of tamoxifen,[7] a compound in which chlorine in MRL-41 is replaced by an ethyl group. The chemists there developed procedures that produced this drug as virtually a single isomer, in which form it is marketed. Extensive clinical trials have shown the drug to be quite efficacious in extending the life of breast cancer patients after surgery. Tamoxifen is, as would be predicted, most effective in patients whose cancer cells are estrogen-receptor positive. The more recently introduced estrogen antagonist toremifene,[8] which has a chlorine atom at the end of the ethyl group, has much the same activity as its predecessors.

Chemists at Upjohn took a somewhat different tack in their search for antifertility agents based on MRL-41. That approach involved analogs in which the two carbons with a double bond were closed back into a ring. The promising activity of the first of those agents, U-11,555 A as an estrogen antagonist was noted in Chapter 6. The somewhat more difficult to prepare congener, with a six-membered fused ring would, they felt, closely resemble the A–B ring portion of the steroidal estrogens. Compounds in this series proved to be almost uniformly more potent than their predecessors with a five-membered fused ring.[9] The contraceptive activity of the most potent of these, nafoxidine (Fig. 6), was also confined to rodents.

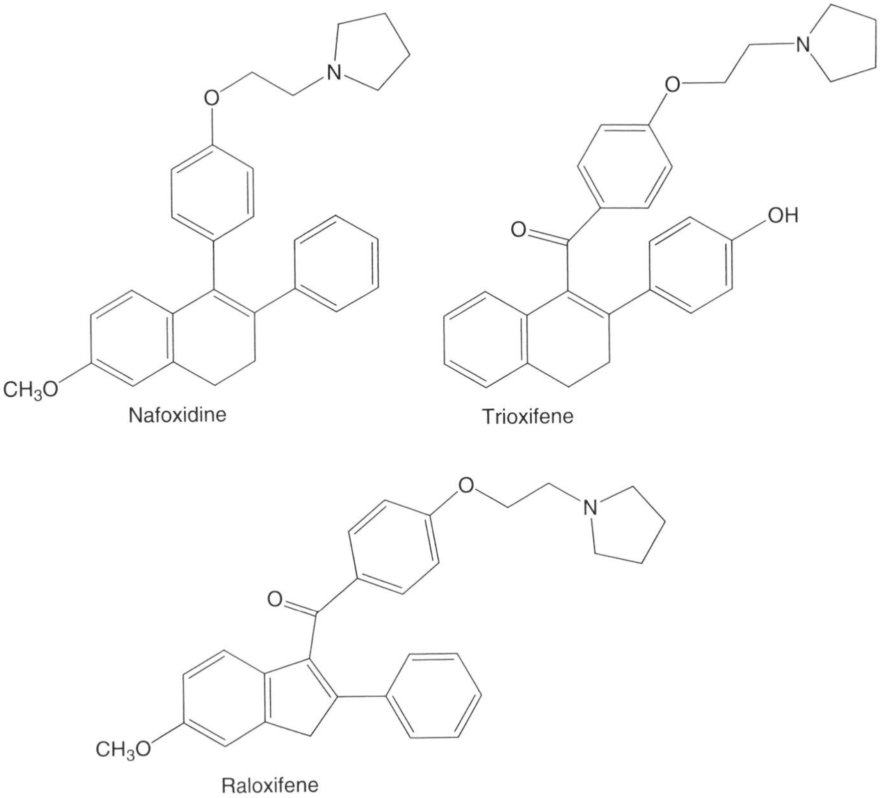

Nafoxidine

Trioxifene

Raloxifene

Figure 6   Fused ring estrogen antagonists.

Although this compound proved useful in treating breast cancer in women, clinical studies were abandoned due to tolerance problems.

Some decades later, scientists at Lilly revisited the field of estrogen-receptor antagonists. The compounds that they investigated differed from the earlier fused ring series in that a ketone (C=O) was inserted between the fused ring and the benzene ring that holds the basic ether. The biological activity of this compound trioxifene at first approximation matches that of the preceding estrogen antagonists. In a subsequent series, the all-carbon fused unit was replaced by a benzothiophene. The sulfur atom in the five-membered ring in these agents occupies approximately the same span as the two carbons in the six-membered ring of the earlier congeners. More detailed pharmacology showed sufficient differences in activity on a number of endpoints to cause compounds in this series to be characterized as "designer estrogens." Many of these agents, for example, retain a higher ratio of estrogenic agonist to antagonist activity than most of the compounds in the earlier series. Raloxifene,[10] the drug that came out of the Lilly series is indicated for treatment of typical estrogen-sensitive endpoints including menopausal and postmenopausal symptoms as well as osteoporosis. The drug does, however, retain some estrogen antagonist activity. It has thus been used in a large-scale clinical trial to test its activity in preventing breast cancer.

The immediate goal of the use of estrogen antagonists in treating breast cancer is to diminish stimulation of estrogen-sensitive tumor cells. An alternative strategy depends on cutting off or at least diminishing the body's own supply of estrogens. This in turn involves blocking the mechanism by which these molecules are synthesized. It had been established as early as the 1950s that cholesterol is the starting material for endogenous steroids such as estrone, progesterone, the other hormone involved in female reproductive function, and its counterpart in the male, testosterone (Fig. 7). In one of a series of enzymatic reactions the side chain at position 17 in cholesterol is completely removed and the alcohol in the A ring oxidized to a ketone. The product of this reaction is androstenedione (androst-4-ene-3,17-dione to a chemist), a compound that still has essentially the same A ring as its precursor, including the methyl group at the juncture of rings A and B. The enzyme aromatase then removes the methyl group and introduces an additional double bond. The overall product of these transformations is estrone. A drug that would inhibit this enzyme would be expected to cut down production of the hormone. The search for agents that would accomplish that task initially focused on modified steroids. This was based on the observation that antagonists often resembled the molecules that were to be blocked. This had proven to be the case in areas as diverse as the adrenergic system (beta-blockers), antibiotics (penicillins and cephalosporins), and of course the estrogen antagonists.

One aromatase inhibitor in fact used in the treatment of breast cancer does consist of a modified steroid. This drug, exemestane (Fig. 7),[11] was approved for human use in the United States in 1999. The chemical structure closely resembles androstenedione, the usual staring material for the enzyme. The resemblance is apparently close enough for it to be accepted. The extra carbon atom with a

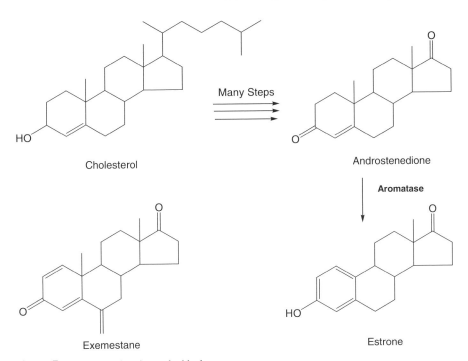

Figure 7   Aromatase function and a blocker.

double bond on the B ring is very reactive and will avidly form chemical bonds with atoms that are rich in electrons. The compound is thus likely to attach irreversibly to the aromatase binding site, in effect closing off the site to normal substrate. This inactivates the enzyme and shuts off synthesis of fresh estrone.

A pair of closely related compounds whose structures differ markedly from steroids also inhibit endogenous steroid production (Fig. 8). The aromatase enzyme, in addition to the site that binds with the substrates, also includes chemical

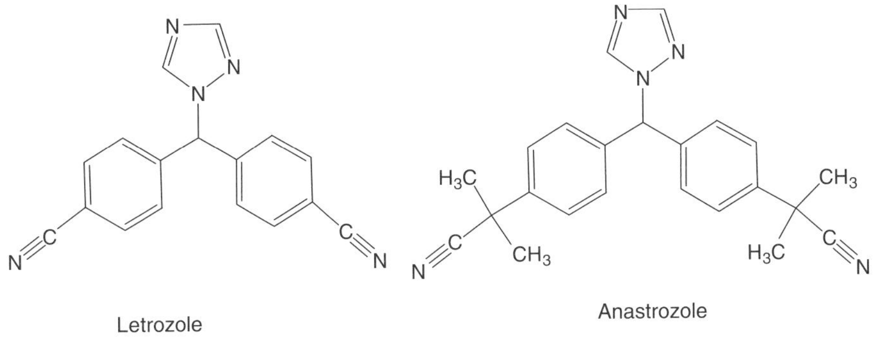

Figure 8   Nonsteroid aromatase inhibitors.

structures involved in its oxidative actions. These parts of the molecule are not strictly speaking receptors, but can and do interact with foreign molecules. The resulting complex can inhibit the enzyme's function. Both anastrotrozole[12,13] and letrozole,[14,15] inhibit estrogen production by binding with that alternative site. These drugs bind in reversible fashion, in contrast to exemestane, which irreversibly inactivates the enzyme. They were approved by the FDA in 1995 and 1997, respectively, for use in breast cancer.

Development of drugs based on the other two classes of hormones involved in reproductive function would prove more difficult. The chemical structures of both testosterone and progesterone are somewhat more complex than those of the estrogens. Research on drugs that acted on estrogen sites, as noted above, profited greatly from the early discovery of relatively simple nonsteroid estrogens. Few, if any, nonsteroid androgens or progestins have been found to date, probably due to the higher specificity of the receptors for the other two classes of the so-called sex hormones. Research was also initially hampered by the fact that starting materials for research on androgen and progestins were available only in minute quantities. A host of natural products that incorporate the steroid nucleus are found in nature. These are but a small part of large number of compounds based on the isoprene unit. Steroid natural products are not suitable for use as such, because they are usually decorated with a large number of substituents that make them unsuitable as starting materials. The sugar molecules attached to many of those products give them soap-like properties due to the combination of the fatty steroid part and the water-soluble sugar. They are thus often called saponins; the portion that remains after the sugar has been stripped off is known as a sapogenin. In the late 1940s the steroid chemist Russell Marker of Penn State launched a program to find sapogenins that could be used for the production of steroid hormones. This culminated with the finding that sapogenins from the root of the Mexican yam could be converted on laboratory scale to progesterone by stepwise removal of a pair of extra oxygen-containing functions attached to ring D. He then undertook a personal search in Mexico for plant sources for those sapogenins. This resulted in the discovery that the root of a wild yam of the Dioscorea species was a rich source of the sapogenin diosgenin (Fig. 9). He then developed a five-step route for converting that chemical to progesterone that could be scaled up for producing the then rare compounds in kilogram amounts. He had little luck, however, in finding a pharmaceutical firm in the United States interested in furthering the work, to say nothing of undertaking commercial production of progesterone. Almost by chance he found a small company in Mexico City that was willing to take on the task. That firm, Syntex, would go on to become one of the leaders in steroids in the coming decades. Within a few years, procedures were developed for using progesterone as a starting material for testosterone and ultimately estrone. Mexican yam roots still comprise an important source for medicinal steroids. It is apparently competitive with the soybean sterols that will be discusses in connection with the corticosteroids.[16]

The menstrual cycle in the human female is marked by a regular rise and fall in blood levels of both estrogens and progesterone. The latter is also intimately

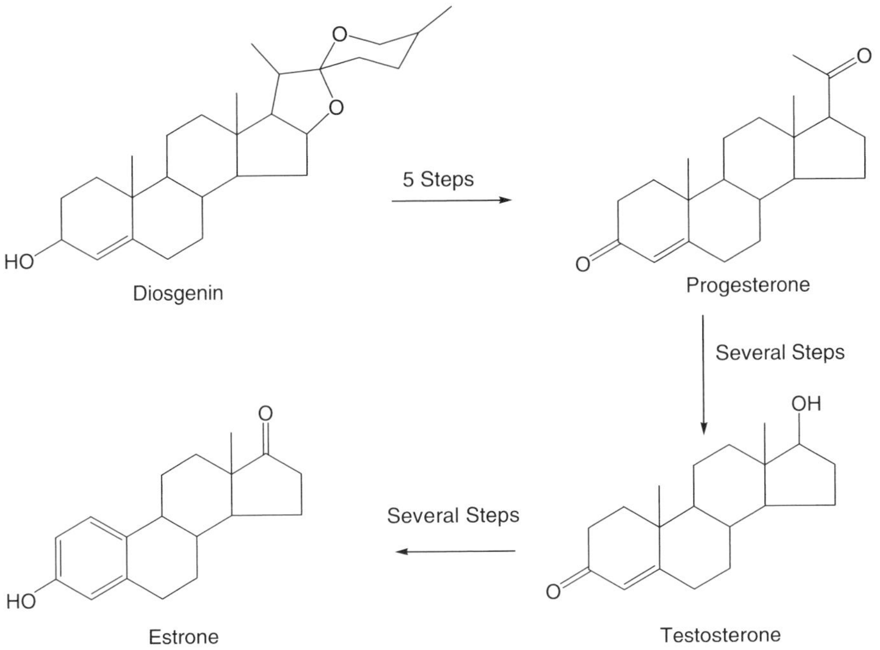

Figure 9   Diosgenin as a source for sex steroids.

involved in supporting pregnancy. As soon as a woman becomes pregnant, the site on the ovary from which the ovum was shed is transformed into a corpus luteum. That structure then becomes an abundant source of progesterone. The resulting high serum levels of hormone help maintain the uterus and not incidentally prevent ovulation. Progesterone supplements would thus, at least in theory, be useful in treating various malfunctions of the reproductive cycle. Progesterone itself, however, must be given by injection, as the drug is not active by the oral route. A considerable amount of research thus went into the search for derivatives that could be given by mouth. This account to some extent departs from strict chronology here as the estrogen-derived progestins discussed further on came first.

Research in various laboratories established that the presence of additional substituents at the 6 and 17 positions increased potency and oral activity. The drugs that came out of these programs all incorporate a hydroxyl group at the 17 position on the five-membered D ring as its ester with acetic acid. The first of these drugs, medroxyprogesterone acetate,[17] which acquired the trade name Provera®, was prepared by chemists at Upjohn (Fig. 10). The compound incorporates a methyl group at the 6 position. Addition of yet another carbon, this as a group connected to the 16 position in the D ring by a double bond, gives the progestin melengestrol acetate.[18] Removal of two hydrogens from ring B in medroxyprogesterone acetate to give a double bond leads to megestrol acetate[19]; replacement of the methyl

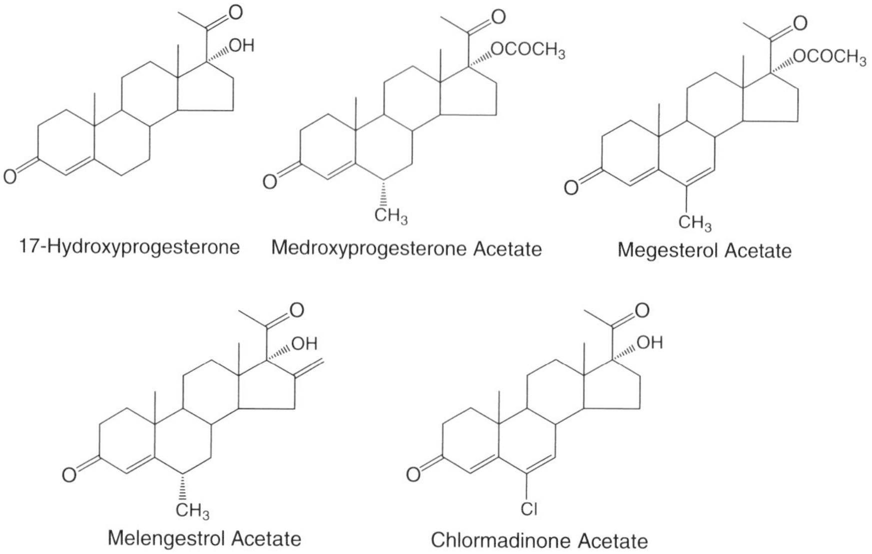

Figure 10   Progestins derived from progesterone.

group at the 6 position in that analog by chlorine leads to chlormadinone acetate.[20] These compounds, as a consequence of their progestational activity, all inhibit ovulation. Their use as oral contraceptives was, not unexpectedly, investigated in some detail. Chlormadinone did prove to be an effective oral contraceptive when administered alone. The drug was, however, withdrawn in 1970 because of an excessive rate of side effects. The high potency of melengestrol, and the low doses required for treatment, led its use in synchronizing estrus in cattle. The drug is particularly valuable in keeping cows out of estrus and thus quiescent during shipment. Provera® is arguably the best known member in this class of compounds. The compound was, for example, at one time used as the progestational component in the oral contraceptive Provest®. This drug was abandoned when it proved noncompetitive with the combinations described further on in this chapter.

Depo-Provera® comprises probably the best-known form of this compound. That product comprises a solution of the drug in an oily medium suitable for injection in a muscle. Drug leaches out of the depot generated at the injection site over a prolonged period. This leads to continuous levels of the progestin in blood. Depo-Provera® has thus proven useful as an injectable contraceptive; a single injection provides coverage for as long as four months. The compound was also used as the progestin in the hormone replacement drug Pempro®. This combination of Premarin® with the progestin was designed to be more "physiological" than unopposed estrogen. Interim analysis of data from a large-scale clinical trial in postmenopausal women revealed an increased risk of invasive breast cancer. The large multicenter trial also found increases in coronary heart disease, stroke, and pulmonary embolism in study participants. This led to an immediate halt of the

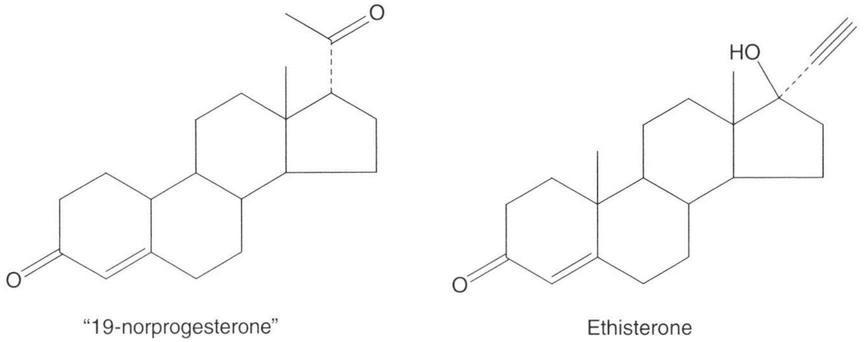

"19-norprogesterone"   Ethisterone

Figure 11 Ehrenstein's "19-norprogesterone" and ethisterone.

trial, an event that generated a flood of coverage in the lay press. The shadow that this event cast on Premarin® markedly decreased sales of that product.

The important role that progesterone plays in the reproductive cycle in the female made this molecule an important focus in the search for oral contraceptives. Support for this came from the observation that progesterone will inhibit induced ovulation in rabbits. The principal drawback to use of this compound as a contraceptive was the fact that it was not active when taken by mouth. Much of the work that led to the first successful oral contraceptives preceded the development of the orally active drugs based on progesterone itself, discussed above, by close to a full decade. Part of the inspiration for this work came from a report in the mid-1940s by Ehrenstein[21] that a progesterone derivative lacking the methyl group at the fusion of rings A and B, the 19-methyl group, showed progestational activity in laboratory animals when administered orally (Fig. 11). This compound, which had been prepared by a lengthy route from the natural product strophanthidin, was later ascertained in fact to be an isomer of progesterone in that the acetyl group ($O=CCH_3$) at position 17 was in the opposite orientation from that in the hormone. The finding that a reversed side chain at the 17 position did not preclude activity on progestin endpoints had in some ways been foreshadowed by some earlier work. In 1937, Ruzicka at the Swiss Federal Technical Institute (ETH) reported on the preparation of a derivative of testosterone that had a two-carbon chain, with a triple bond, in the reversed orientation from progesterone.[22] This compound, ethisterone, showed modest progestational activity when administered orally.

The key to providing a practical route to compounds lacking the 19-methyl group was the eponymous reduction procedure developed by A. J. Birch at the University of Sydney.[23] As later modified by others, the procedure in essence comprises treating the methyl ether of a phenol with a solution in liquid ammonia of an active metal such as lithium or sodium. The resulting product is an enol ether that contains two more hydrogens than the starting material. This modification eliminates the benzenoid properties of the ring. Treatment of such compounds with mild acid converts them in two steps to an unsaturated ketone. When applied to a

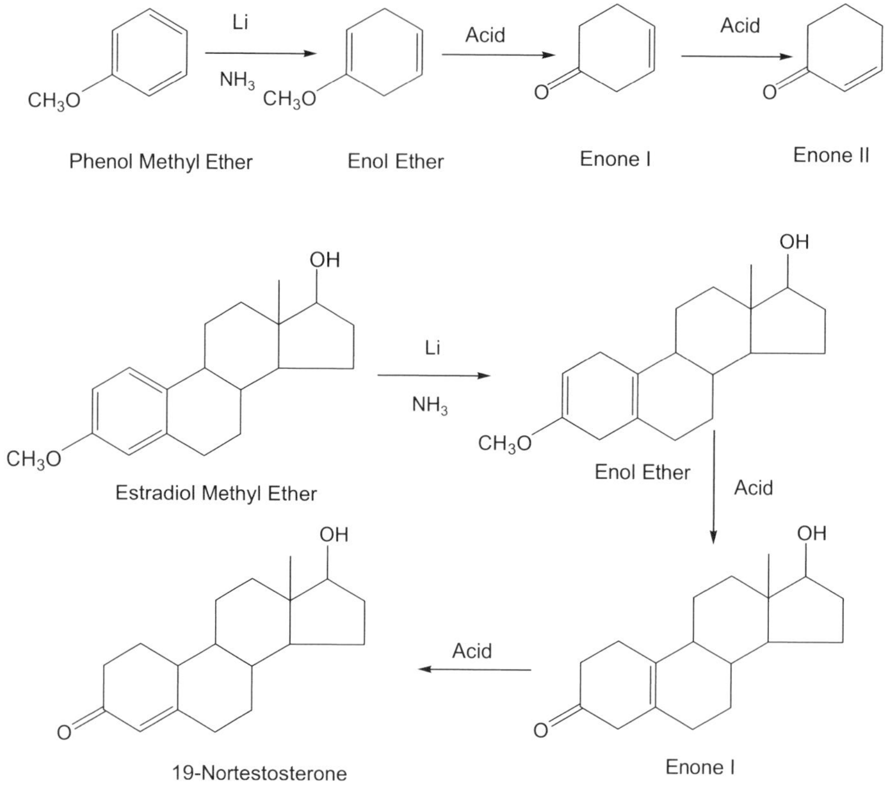

Phenol Methyl Ether   Enol Ether   Enone I   Enone II

Figure 12   The Birch reduction and its application to a steroid.

steroid with a benzene A ring such as the methyl ether of estradiol, the end product in effect consists of testosterone lacking the 19-methyl group at the 10 position at one of the junctions of rings A and B. This compound is usually designated by its nonsystematic name 19-nortestosterone. The Birch reduction was to provide the key to a very large class of compounds known as the 19-norsteroids (Fig. 12).

It might be noted in passing that some mystery is attached to the origin of the widely used prefix "nor" first used for adrenergic hormones as in, for example, noradrenaline. This modifier derives from neither the Greek or Latin term for absence. An apocryphal, but credible, account traces back to some unknown laboratory in Germany where chemists were preparing a variety of closely related analogs by hooking various side chains onto a common nitrogen-containing core. The use of the generic designation R for side chains was by then well established among organic chemists. The bottle of nitrogen-containing core is said to have borne a label that read "N Öhne R" (N without R), a label that was later shortened to NOR.

The accumulated biological data, as well as the newly developed chemistry, pointed to an approach for preparing orally active progestins that would inhibit ovulation in women. Theory indicated that such a product would finally yield the

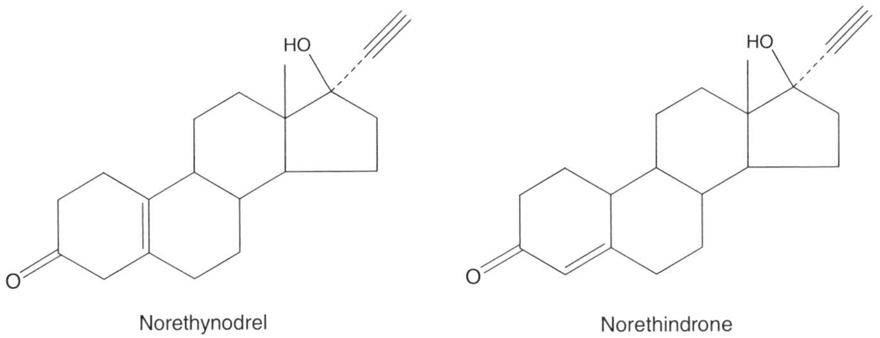

Figure 13 Norethynodrel and norethindrone, the first oral contraceptives.

long sought oral contraceptive. As has often been the case, two different groups independently came up with almost identical molecules (Fig. 13). Frank Colton and his colleagues at G. D. Searle in Chicago prepared the 19-nor steroid norethynodrel. This compound can be viewed as a version of ethisterone in which the 19-methyl group at the junction of rings A and B has been omitted. It differs from the model, however, because the method of synthesis leaves a double bond in an unusual position at the A–B ring fusion. After the drug had cleared safety testing it was subjected to large-scale trials as a contraceptive. Results from those trials confirmed that the drug was indeed an effective contraceptive. The compound, by now named Enovid®, was initially approved in the United States in 1962 as an orally active progestin. Approval as an oral contraceptive followed not much later. This drug and its successors were soon to be known as simply "The Pill."

At about the same time, Carl Djerassi and his associates at Syntex announced the preparation of another orally active 19-nor progestin.[25] This compound, norethindrone, bears a closer relation to ethisterone, because the double bond in the A ring is now in the same position. As chemical theory would predict, exposure of norethynodrel to strong acid will quickly convert the compound to norethindrone. The likelihood that this happened in the stomach has led to warm exchanges as to the priority of the invention of the Pill.

The majority of current oral contraceptive drugs comprise a combination of a 19-nor progestin with a minute amount of an estrogen. An unverifiable, though credible, account attributes the discovery of the need for that second ingredient to the unexpectedly high rate of pregnancies observed during early trials of the orally active progestin ethynodiol diacetate. The state of analytical chemistry when this compound was developed helped ensure that the drug substance used for those trials was unusually pure. The treatment failures with ethynodiol diacetate led to examination of Enovid® using the newly developed analytical methods. This revealed that what had been thought to be a pure compound was actually contaminated with a small amount of the methyl ether of 17-ethynyl estradiol, a compound later assigned the nonproprietary name mestranol (Fig. 14). Its presence in the final product traces back to the fact that organic reactions more often than not do not go

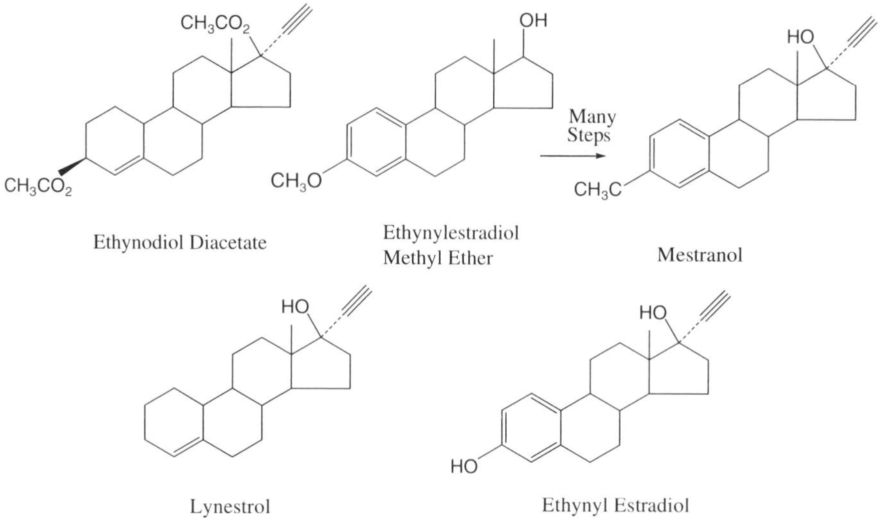

Figure 14   Ethynodiol diacetate and mestranol.

to completion. A small amount of estradiol methyl ether thus survives the Birch reduction of estradiol methyl ether (see Fig. 12). This contaminant is then converted to mestranol when exposed to the subsequent set of reactions that lead to ethynodrel. The amount of this estrogen was carefully adjusted to a specified amount in all subsequent batches of Enovid®. Ethynodiol was then reformulated to include a small amount of mestranol. Data from additional biological studies then led to the conclusion that the estrogen component played a role in the inhibition of ovulation induced by the oral contraceptives. The large market for these drugs that emerged in a very short time for these drugs sparked research in other firms to develop their own drugs. The structure of oral progestin lynestrol,[26] for example, seems to suggest that omission of the oxygen atom at the 3-position is compatible with hormonal activity. The proportion of estrogen and even the overall dose were considerably reduced for most oral contraceptives over the coming years. The estrogenic component, mestranol, has been replaced by ethynyl estradiol in most products. The ratio of progestin to estrogen among various oral contraceptives varies over the range from 10:1 to 50:1.

The estradiol starting material used for the original 19-norprogestins was derived from androstenedione, which in turn came from progesterone. The first step involved conversion of the A ring to its fully unsaturated benzenoid form by extruding the 19-methyl group. In much of the early work, this transformation involved drizzling a mineral oil solution of the starting steroid over glass beads heated to 600°C, conditions that were strenuous enough to crack some of the solvent to gasoline. The development of a chemical procedure for the same task greatly eased access to 19-nor steroids.[27] This problem was completely bypassed after chemists developed a route for synthesizing steroids with a benzene-like A ring

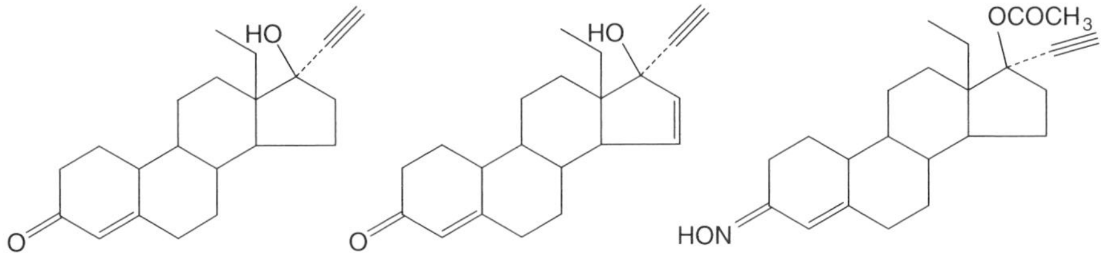

Figure 15   18-Ethyl progestins from total synthesis.

by total synthesis from so-called coal-tar, or more accurately petroleum-derived starting chemicals; this technology in addition allowed chemists to produce 19-nor steroids with major structural changes that could not be realized by modifying natural-steroid-derived compounds. The achievement of such a method for the total synthesis of estrone arose in independent reports by Ivan Torgov of the Zelinski Institute in Moscow[29] and Herschel Smith of Wyeth.[28] Using this route, the latter went on to prepare an analog of ethindrone where the 18-methyl group at the juncture of rings C and D was replaced by ethyl. The Torgov–Smith synthesis is not stereoselective; that is to say that it yields a product that consists of exactly equal amounts of the two mirror image forms, a so-called racemate. This product, later known as norgestrel,[30] was formulated as an oral contraceptive without added estrogen (Fig. 15). The drug been largely replaced by the single mirror image form levonorgestrel. High single doses of this compound have been found to be a quite effective postcoital contraceptive when used within a short time after the event. This product is now available as the prescription drug Plan-B®. Attempts by Barr Laboratories to gain FDA approval for over-the-counter sales have led the agency to bow to political expediency by deferring action ad infinitum.

The Torgov–Smith synthesis provided a relatively ready route for the preparation of other analogs that incorporated the unnatural 18-ethyl group. A number of these compounds are now in clinical use as oral contraceptives in combination with small amounts of ethynyl estradiol. Gestodene[31] incorporates a double bond in the five-membered ring, a feature not found in any of the natural steroids. The progestin norgestimate[32] may be viewed as a derivative of norgestrel in which the alcohol at 17 has been converted to an ester and the ketone at the 3-position to an oxime.

Several additional drugs in the 19-nor series are worth noting. The unsaturated derivative dienogest[33] contains double bonds in both the A and B rings (Fig. 16). This oral contraceptive differs from its predecessors further in that the usual triple-bonded acetylene function at the 17-position is replaced by a cyanoethyl group. The presence of the ethyl group in gestrinone[34] might at first sight suggest that it too derives from a Torgov–Smith intermediate. This analog is, however, prepared by a novel synthetic route developed by chemists at Roussel–UCLAF. The scheme developed by Veluz and his associates differs from previous approaches mainly in the fact that it does not at any stage involve an intermediate with an

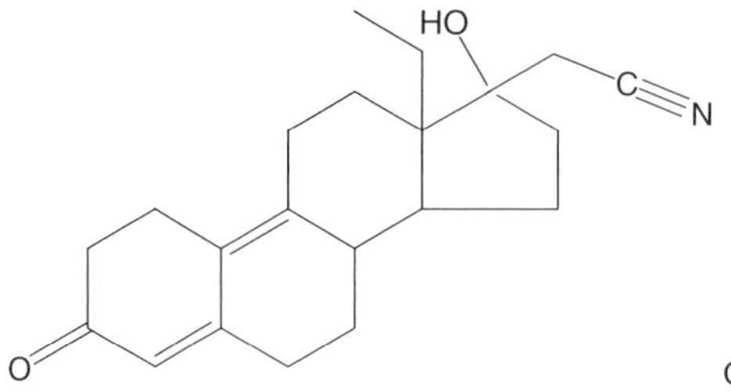

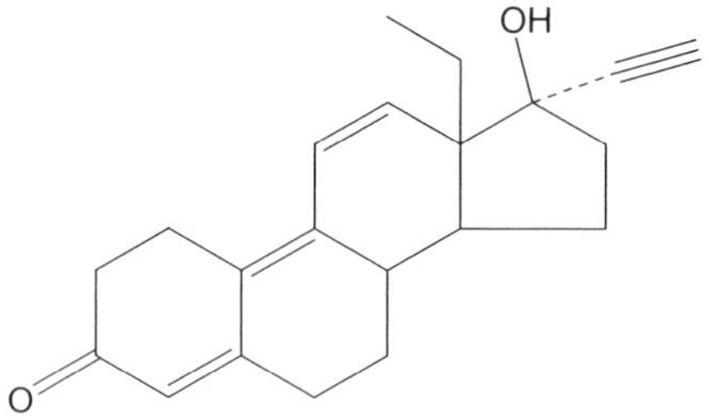

Dienogest   Gestrinone

Figure 16   Unsaturated 19-nor progestins.

estrone-like A ring. Bypassing such an intermediate and the necessity for the Birch reduction obviates the possibility of contamination with estrogenic impurities in the final product. This new route also provided the chemical basis for the preparation of progestin antagonists. The very highly unsaturated compound gestrinone differs further from its close relatives in that it is indicated for use as a progestin rather than as a contraceptive.

The flexibility of this new scheme for the preparation of 19-nor compounds by total synthesis and the resulting availability of intermediates with unusual structures led the chemists to undertake wide-ranging projects to explore the effects of major modifications. It was well known at the time that the presence of substituents at the 11-position in the C ring was, as will be noted later, crucial for the activity of corticosteroids, also called corticoids. The new chemistry afforded a relatively convenient method for preparing compounds that carried unusual chemical groups at that position that differed radically from those found in natural hormones. G. Teustch and his associates at Roussel–UCLAF prepared a number of derivatives that included variously modified benzene rings at the 11-position. They were encouraged in this work by the finding that some of the early examples antagonized the effects of corticoids. Specific analogs, however, seemed also to antagonize the effects of progestins. A specific progesterone antagonist would, among other effects, potentially reverse one its principal roles, that is, providing hormonal support for the pregnant uterus. One of the compounds from this program, mifepristone, far better known by its code number RU-486, was shown to act as a specific antagonist at progesterone receptors, but the compound had little anticorticoid activity (Fig. 17). Clinical testing confirmed that this drug was a very effective abortifacient. The prostaglandin derivative, misoprostone (see Chapter 6), is often administered in conjunction with mifepristone to clear the uterus. The heated politics that accompanied (and still do) approval and market introduction in the United States are beyond the purvey of this volume.

The orientation of the methyl group at the C–D ring junction, as well as the groups at the 17-position are interestingly reversed from those in mifepristone in the analog from Schering (Berlin), onapristone.[36] This change in orientation does not affect activity, as this too is an effective progesterone antagonist. The

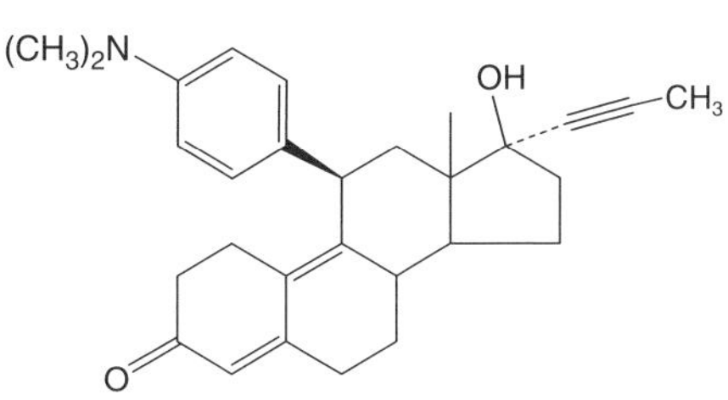

Mifepristone

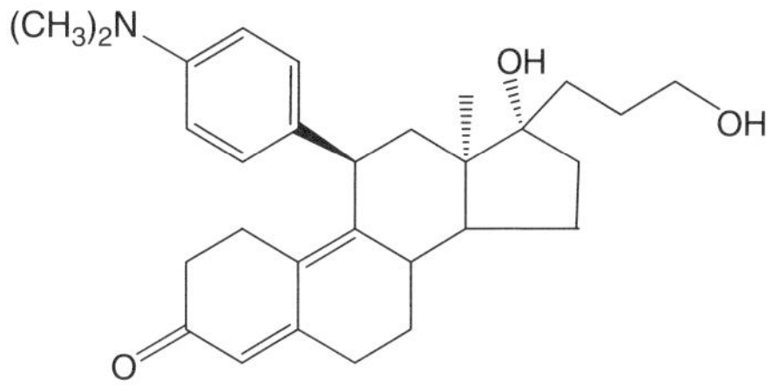

Onapristone

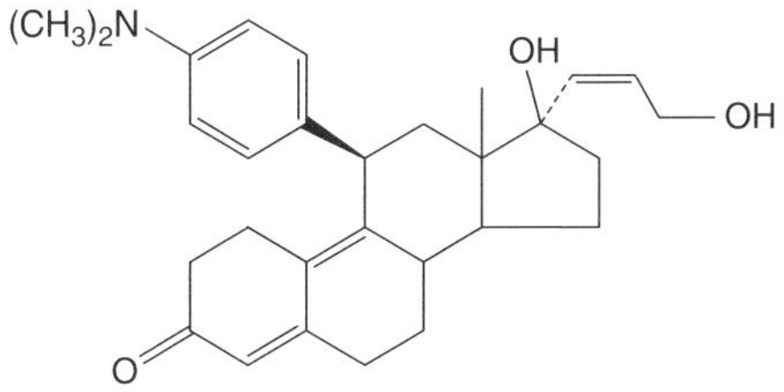

Lilopristone

Figure 17   Progesterone antagonists.

drug was, however, withdrawn from the clinic due to liver toxicity. The analog lilopristone[37] has those groups in the normal orientation and an additional double bond in the hydroxyl-terminated side chain.

Testosterone and its structural congeners, known by the collective name androgens, play an analogous role in male reproductive functions to that of the estrogens in the female of the species. Testosterone and its congeners thus support the formation of sperm and development and later support of male sex organs and secondary sex characteristics. There is thus a readily apparent need for androgens as drugs for replacement therapy for those individuals who, for any one of a number of reasons, suffer from deficient androgen levels.

Another important property of the androgens is their role in stimulating the development of muscle mass, the so-called anabolic effect. Some athletes began consuming these drugs starting in the 1950s, almost as soon as androgens became available as prescription drugs. Weight-lifters, whose performance relies most directly on massive muscles, may well have led the way to relying on the anabolic effect of androgens for gaining advantage over their less pharmacologically sophisticated competitors. The presumed edge provided by these drugs to athletic performance led to their surreptitious adoption in one sport after another. Steroid abuse eventually spread across age groups to encompass even aspiring athletes in high schools and, it is said, grade schools. Androgen consumption for this sports application, not surprisingly, far outweighs that for replacement therapy. The Controlled Substances Act passed in 1970 was targeted almost exclusively at drugs that act on the central nervous system and had at least some potential for causing addiction. That law set up categories ranging from V to I, which imposed increasingly stringent controls on prescribing physicians and dispensers. The concern raised by the ever-spreading use of anabolic agents caused those compounds to be added in 1990 to the list of controlled substances, despite the fact that they are not addicting in the usual meaning of that term. The listing of these compounds as Schedule III drugs is the second-most stringent for agents that have a recognized therapeutic use. The success of these regulations must be set against the perennial scandals involving the revelation that some famous athlete has tested positive for anabolic compounds. These compounds are most often described in a condensed manner in the press as "steroids." The unwavering demand for these drugs has also set off an ongoing contest between analysts and chemists in clandestine laboratories. The latter assiduously peruse the literature for reports of androgens with unusual chemical structures in the hope that those will not be detected by existing assays for anabolic steroids. This is a challenging task as fifty steroids are to be found in the current Schedule III list. The selection of compounds covered in this account is thus far from exhaustive. All are, of course, considered to be controlled substances.

Most of the drug development research on androgens was carried out in the late 1950s and early 1960s. The signature male hormone, testosterone, (Fig. 18) is very poorly available when taken by mouth. It might be added parenthetically

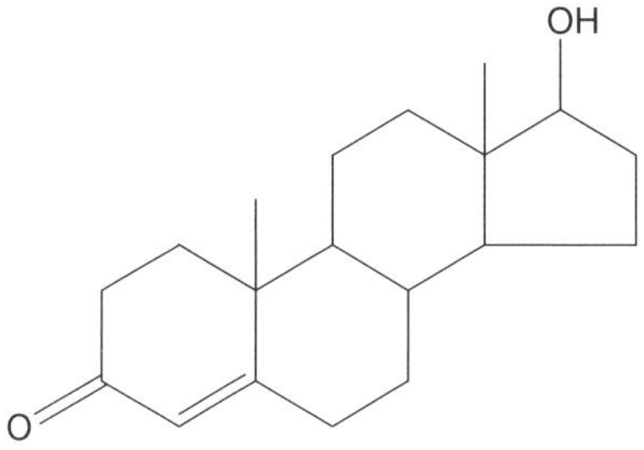

Testosterone

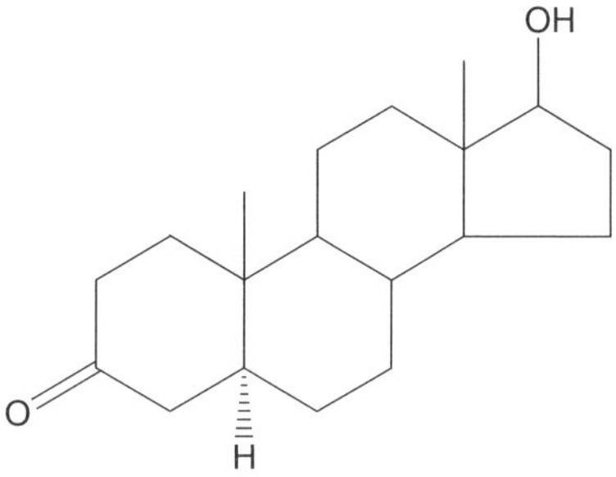

5α-Dihydrotestosterone

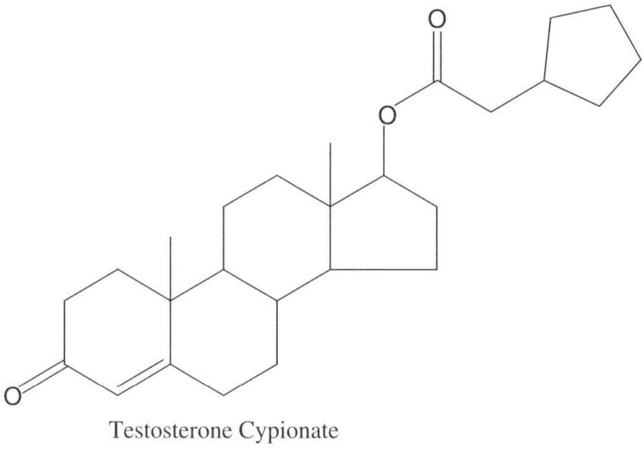

Testosterone Cypionate

Figure 18   Testosterone and related androgens.

that the amount that gets into the bloodstream is not the ultimate active agent. The compound that binds at the receptor level is actually 5α-dihydrotestosterone, which is formed in the liver. Early attempts to get around the problem of lack of oral activity involved the preparation of very slowly hydrolysed esters such as testosterone cypionate. These were dissolved in oil and administered by injection to form a depot. The drug leaches into the bloodstream where circulating ester cleaving enzymes will free the parent drug and provide constant levels of testosterone.

Lack of oral activity was traced to very fast inactivation of the alcohol at position 17 by oxidation in the liver. Adding a methyl group to that position in principle and in fact precludes that process. The product of that modification, 17-methyl testosterone, prepared in 1937 in the early days of steroid research, was in fact active when administered orally. Several more such relatively simple compounds related to testosterone that depend on a methyl group for protection of the 17-position were introduced by the early 1950s. The first of the agents comprises the anabolic androgenic agent methandriol,[38] first prepared in 1935, although possibly not tested for biological activity at that time. Other examples include methandrostenolone[39] and oxymestrone (Fig. 19).[40]

Additional modification of the basic structure increased potency. Adding an additional nitrogen-containing ring gives stanazole[41]; oxandolone[42] is a

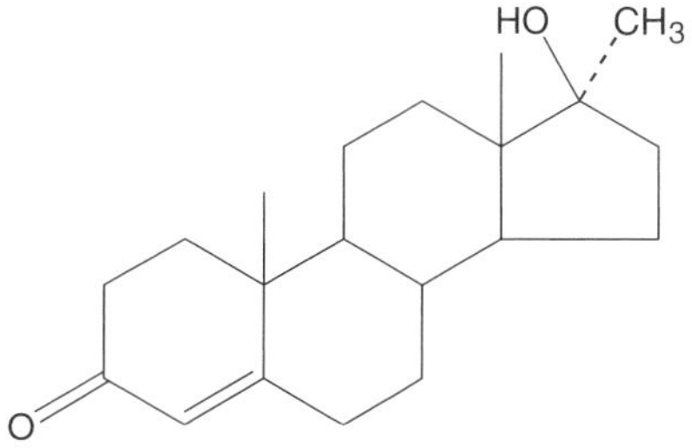

17-Methyl Testosterone

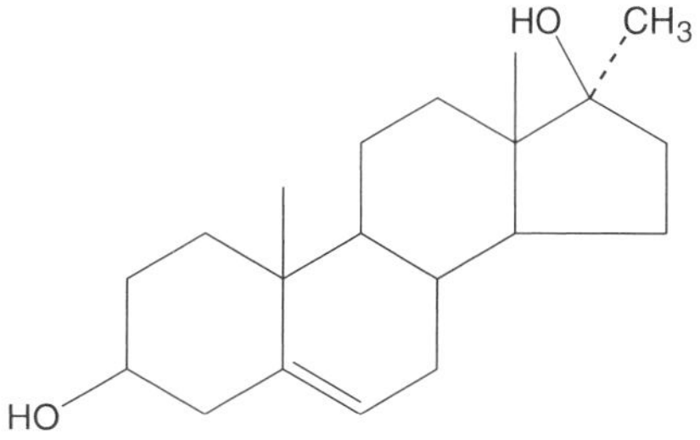

Methandriol

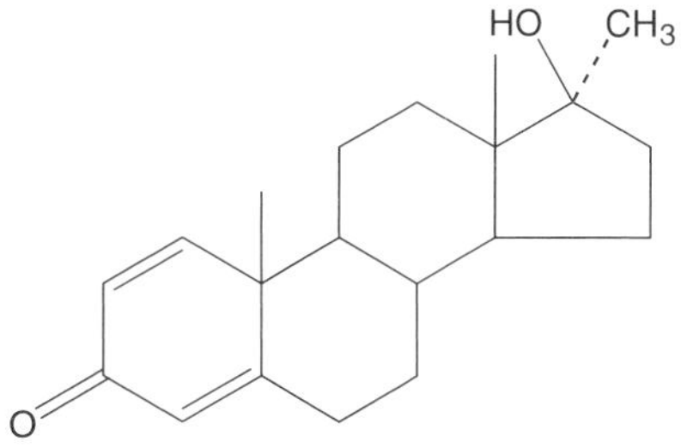

Methandrostenolone

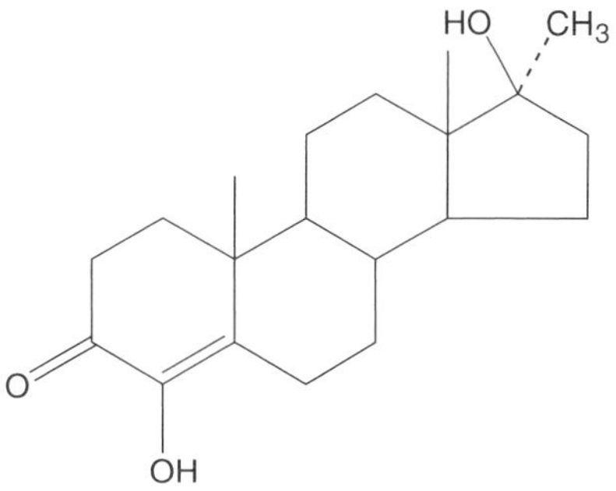

Oxymestrone

Figure 19 17-Alkyl androgens.

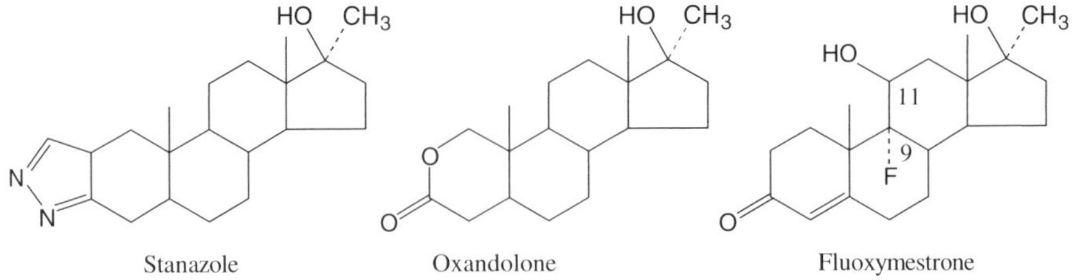

Figure 20   More complex 1-alkyl androgens.

compound in which one of the carbon atoms in ring A of testosterone is replaced by oxygen (Fig. 20). The rather more complex androgen fluoxymestrone[43] contains an alcohol at the 11-position and fluorine on carbon 9. This combination of modification, which is more characteristic of the corticosteroids, discussed further on, increases potency. Synthesis of this compound in fact starts with an intermediate used in the preparation of analogs of cortisone. Data from longer term usage of many these drugs showed a shared propensity to cause liver damage. It became lore among workers in this field that all testosterone derivatives that had hydrocarbon groups at 17 would share this effect.

The chemistry that led to the 19-norprogestins provided the basis for several androgenic steroids as well. Early results from animal studies suggested that some of the agents showed preferential anabolic over androgenic activity. This held out the promise of drugs that could be used to treat wasting diseases without concomitant masculinizing effects. These apparent splits in activity did not hold up in clinical usage. The first compound in this series, nandrolone,[44] is simply the product obtained by the Birch reduction of mestranol (Fig. 21). The structure of this compound in effect consists of testosterone that is missing the methyl group at the 10-position. This compound, which could also be called 19-nortestosterone, shows the same activity as the natural hormone. It is also, like its counterpart, quickly destroyed by oxidation in the liver of the 17-hydroxy group. The compound is thus used in the form of an ester with a fatty acid such as decanoic acid. Nandrolone decanoate is formulated in oil for use in long-acting depot injections. The analog, norethandrolone,[45] is prepared using methodology that is analogous to that used to prepare some of the early 19-norprogestins. The compound differs from the progestin norethindrone only in that the side chain at the 17-position is a fully hydrogenated ethyl group instead of a triple-bonded acetylene. The compound shows androgenic and anabolic rather than progestational activity in spite of that apparently minute difference in structure. The more recent anabolic agent, trenbolone, is derived from intermediates prepared by the alternative route to 19-nor compounds that does not go through estrogen intermediates.[34] This very potent compound is apparently quite active orally, in spite of the presence of a readily oxidizable alcohol at the 17-position. This potent anabolic agent is one of the so-called designer drugs so highly favored by body builders.

As noted before, one of the key roles of androgens is the support of the male reproductive function. The prostate gland is thus very responsive to androgens.

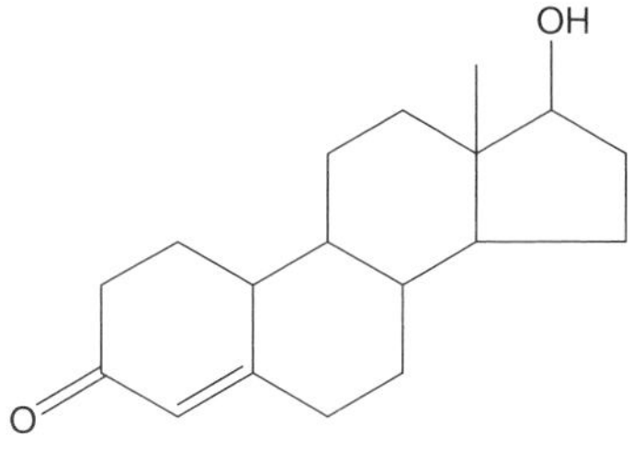

Nandrolone

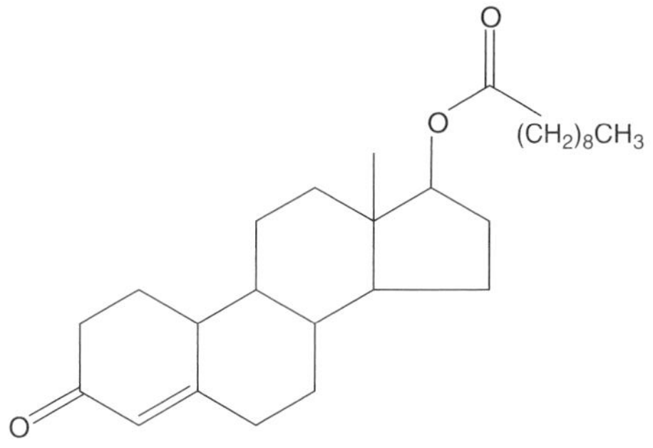

Nandrolone Decanoate

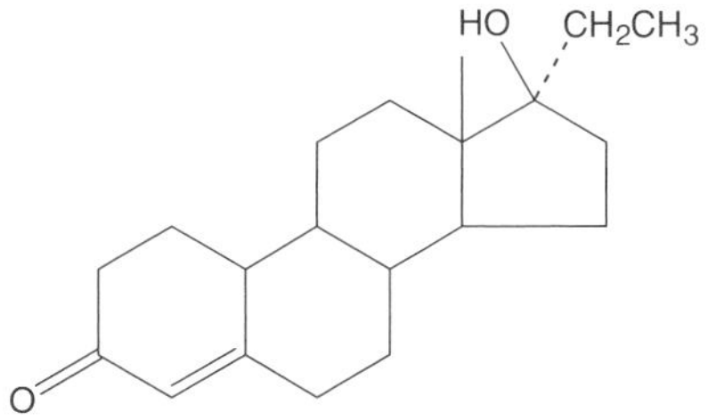

Norethandrolone

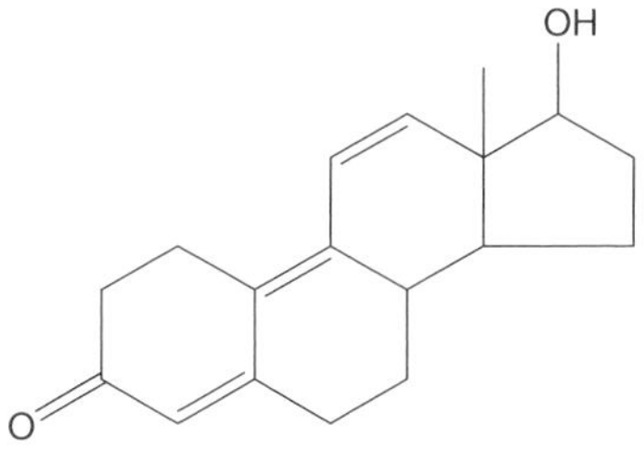

Trenbolone

Figure 21 19-Nor androgens.

The overgrowth of this gland experienced by many men as they age often leads to adverse consequences. The minimal effect, benign prostatic hypertrophy (BPH), results in a variety of annoying effects caused largely by constriction of the urinary urethra by the enlarged gland. The far more serious consequence, prostatic cancer, is the major neoplasm found among aging men. Stimulation of the prostate by androgens and, more specifically dihydrotestosterone, the active metabolite of testosterone, plays a role in both conditions. The search for an androgen antagonist that would block or at least diminish that stimulus dates back to the late 1960s. The finding that the estrogen antagonists identified at that time were useful in treating disease that involved estrogen-dependent tissues showed that this was a viable approach. As noted previously, discovery of the estrogen antagonists was aided by the availability of nonsteroidal estrogens. Research on androgen antagonists was made more difficult by the fact that no nonsteroid androgen had yet been identified; the full steroid nucleus seemed to be required for androgenic activity.

The first androgen antagonist was in fact based on a compound that included a major portion of the steroid nucleus (Fig. 22). This compound, finasteride, reported in 1986 by a group from Merck,[46] did however incorporate an important change from natural androgens. Replacement in this compound of the carbon atom at the 4-position by nitrogen transforms that ring into a cyclic amide. The binding affinity of that ring will differ from that of a normal all-carbon cyclic ketone. Finasteride is not an androgen antagonist by the strictest definition of the term, which requires

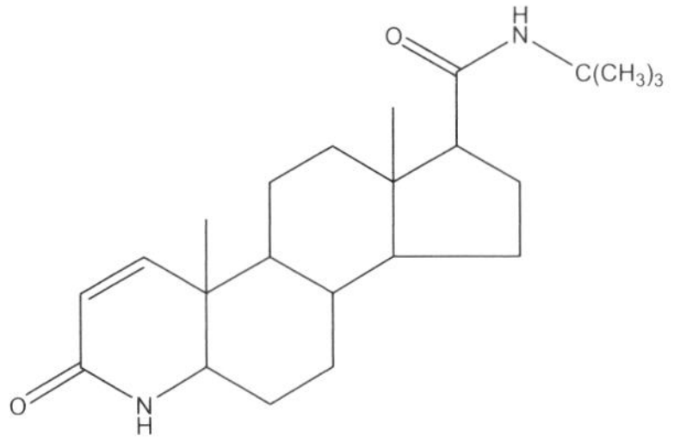

Finasteride

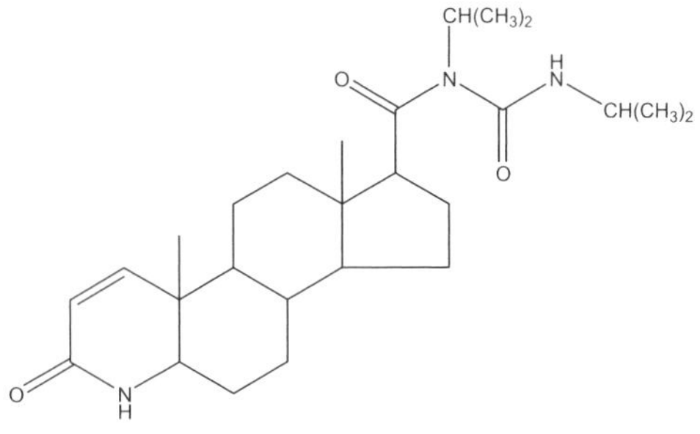

Turosteride

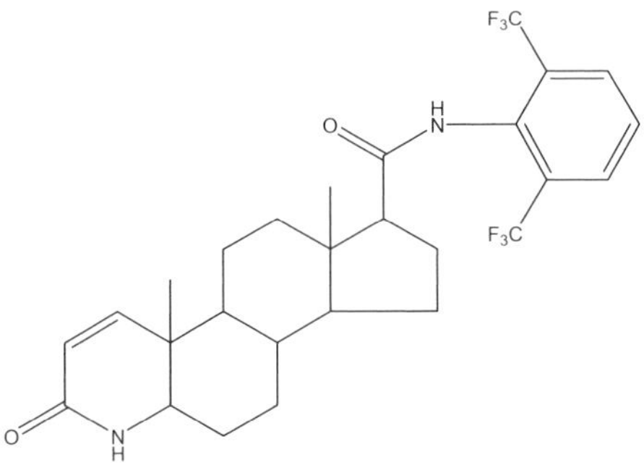

Dutasteride

Figure 22 5-alpha reductase inhibitors androgen antagonists.

competitive binding to androgen receptors. The compound instead binds to and inhibits the enzyme 5-alpha testosterone reductase. This in essence prevents conversion of testosterone to its active metabolite. The drug has proven useful for treating BPH, in many cases shrinking enlarged prostates. A large-scale clinical trial, showed that the drug reduced the risk of developing prostate cancer by 25%.[47] Reductase inhibitors do not, however, seem to be used for treating existing prostate cancer.

Male pattern baldness accounts for a large proportion of men who lose their hair with advancing years. It has been recognized for many years that this cosmetic problem can be traced to circulating androgens. Clinical trials on a formulation of finasteride showed that the drug restored hair in a selected group of subjects. The drug was approved for that indication in 1997. Propecia®, the trade name for this formulation of finasteride, is ingested as a tablet, in contrast to minoxidil, which must be administered as a topical lotion. The simple amide at the 17-position in this drug is replaced by a urea that contains an additional nitrogen atom in the reductase inhibitor turosteride.[48] The amine in the amide at 17 is replaced by an aniline where nitrogen is flanked by trifluoromethyl groups (—CF_3) in the more recent analog dutasteride. This compound is reported to inhibit both forms of the reductase enzyme.[49]

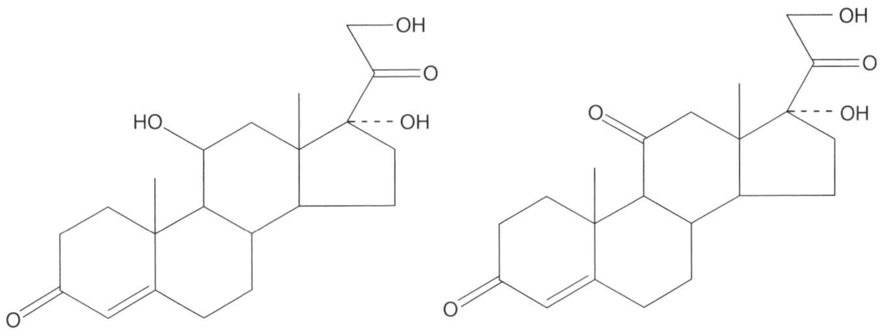

Figure 23   Cortisol and cortisone.

Corticosteroids encompass the third broad class of hormones based on the steroid nucleus. These compounds, often simply called corticoids, are synthesized from progesterone in the adrenal cortex, hence their name. The main product from that source is cortisol, which maintains homeostasis by promoting increased serum glucose levels via breakdown of lipids and proteins. This action is particularly prominent in reaction to stress. Cortisol, its immediate precursor, cortisone, and the various synthetic congeners are, as a consequence, classed collectively as glucocorticoids (Fig. 23). The corticoid aldosterone, discussed in Chapter 3, regulates the balance of inorganic salts in serum, leading to it being classed as a mineralocorticoid. It should, however, be noted that glucocorticoids also have some effect on salt balance. This ancillary activity was to prove a particular problem with some of the potent synthetic analogs, as this led to a host of serious side effects.

One of the more immediately apparent applications for cortisol, as with the other steroid hormones, would involve treatment of patients with deficient levels of that hormone. It was known, for example, by the early twentieth century that Addison's disease was due to deficient levels of an as yet uncharacterized substance from the adrenal gland. It had in addition been shown by 1930 that an extract from the adrenal gland would prolong the lifespan of dogs whose adrenals had been removed. Three groups, Edward Kendall at the Mayo Clinic, Thadeus Reichstein at CIBA and the ETH, and Oscar Wintersteiner at Columbia University undertook independent research projects to identify and determine the structure of the compounds responsible for that activity. Extraction from beef adrenals and painstaking separation led them to identify a number of closely related steroids that clearly differed from the by then well-known sex hormones. Cortisol proper was isolated in pure form in 1937 and its chemical structure established by the early 1940s. The difficulty of the work may be judged from the fact that it took 1000 pounds of beef adrenals to obtain 333 mg of cortisol.[50] By the late 1940s cortisol had been accumulated in gram quantities at Mayo.

A physician at the institution, Philip Hench, had noted that the severity of arthritis was much reduced in women during pregnancy. It was by then known that cortisol levels are elevated during this stressful time. This raised the possibility that relief from arthritis pains might be a consequence of those high levels of cortisol. In order to follow up on that, Hench and Kendall then collaborated on a trial to evaluate the effect of administering the pure compound to arthritic patients. The subjects who received the drug reported almost immediate relief of their symptoms.[51] Reports of that finding launched a concerted effort in industry to develop cortisone and hydrocortisone as drugs. In 1950 Hench, Kendall, and Reichstein shared the Nobel Prize for Physiology for their discovery of the corticoids.

The chemical structures of the corticoids include some unusual structural features. These include the extra hydroxyl groups on the side chain on the 17-position and most importantly the oxygen atom at position 11 in ring C. This occurs as an alcohol in cortisol, more commonly known as hydrocortisone, and a ketone in cortisone, a pair of compounds that readily interconvert in the body.

The oxygen atom at the 11-position posed a large hurdle to any scheme for preparing these compounds in quantity. The corticoids from mammals were at the time the only steroids that included that particular structural feature. The isolated nature of the 11-position precluded any thought of adding the oxygen by some purely chemical operation on progesterone. Rather than rely on chemical synthesis, scientists at G. D. Searle decided to rely on the process used by nature in order to accumulate supplies of cortisone. They accordingly developed a process for producing the drug by passing progesterone solutions though massive columns packed with beef adrenals. Chemists at Merck took a purely synthetic approach to the problem. Their scheme started with a compound that had oxygen at the 12 position, one ring atom removed from their target. The starting material, cholic acid, is one of the bile acids abundantly available from slaughterhouses (Fig. 24). The arduous scheme involved first degrading the side chain at position 17 to the highly oxygenated two-carbon unit found in cortisone, transposing the 12 oxygen to the 11-position and finally adjusting the functional groups in

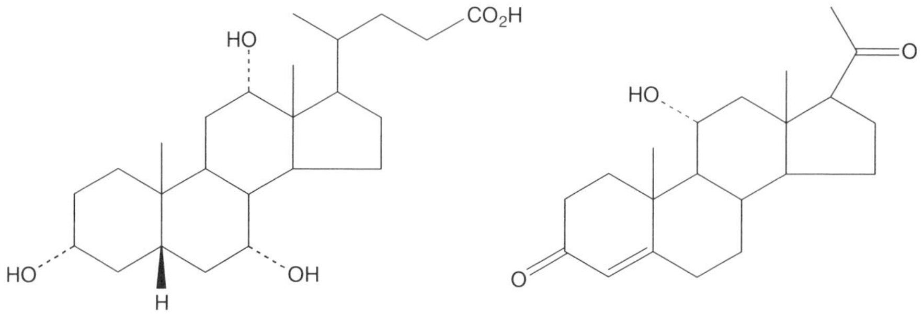

Cholic Acid   11α-Hydroxyprogesterone

Figure 24   Cholic acid and 11α-hydroxyprogesterone.

rings A and B. The published 36-step procedure[52] was later considerably improved and shortened.

Scientists at Upjohn, on the other hand, undertook a program to find a source in nature that would perform the same oxidation step as the enzyme found in adrenal glands. Based on the laboratory's strength in microbiology, they fermented progesterone with a large number of fungi and actinomcytes. The group, led by Dury Peterson, identified several organisms in the *Rhyzopus* family that would introduce an oxygen atom at the 11-position of progesterone.[53] Although the hydroxyl group is on the back instead of the front face of the molecule, as in dihydrocortisone, simple oxidation converts that to the ketone, which can be converted by reduction to the alcohol with the correct orientation. The announcement of this discovery is said to have led to the precipitate abandonment at Searle of the perfusion route, which had already accumulated a significant supply of cortisone. The bile acid route was probably also dropped not too much later. Chemists at Upjohn then developed an efficient route for introducing the additional hydroxyl groups at the 17-position and at the end of the side chain.[54] This new chemistry, in combination with the new 11-hydroxylation procedure, provided an efficient method for producing corticoids.

The world supply of progesterone starting material was at that time largely dedicated to the production of 19-nor steroids and androgens. The humble soybean provided an alternative source due in good part to the work of Percy Julian. This pioneering African-American research chemist, a rarity in the 1930s, had already established his reputation while at De Pauw University with the first total synthesis of the alkaloid physostigmine. He moved on from there to become head of the Soya Products Division of the Glidden Company. One day he was presented with a white deposit that had formed at the bottom of one of the soybean oil tanks that had developed a water leak. He identified that powder as a mixture of steroids that contained significant amounts of sitosterol and stigmasterol (Fig. 25). This accidental finding was soon developed into a systematic procedure for concentrating the small amount of plant sterols present in soybean oil. These soy sterols were to become a major product for Glidden.

The double bond in the side chain of stigmasterol provides a reactive point for removing the side chain. This compound had actually been taken all the way to progesterone as far back as 1934.[55] This led Upjohn chemists to investigate soybean sterols as raw material for cortisone. However, the lack of such a weak link makes sitosterol quite useless as an intermediate. The first step thus involves separating stigmasterol from the mixture. This was initially accomplished by column chromatography; the mixture was absorbed on a column packed with a support such as silica and a carefully chosen solvent percolated through. The different components of the mixture would then, in a best-case scenario, elute in sequence. By systematically investigating the solvent mixtures as the laboratory gained experience, the amount of support was reduced. This was eventually brought to zero as the method turned into a simple process where stigmasterol was leached from the mixture. A small amount of the left-over sitosterol was sold to firms that marketed the product as a cholesterol-lowering agent. Huge cakes of this remainder were

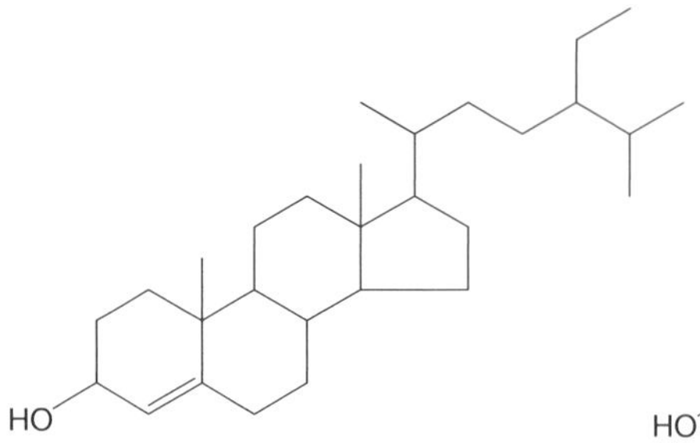

Sitosterol

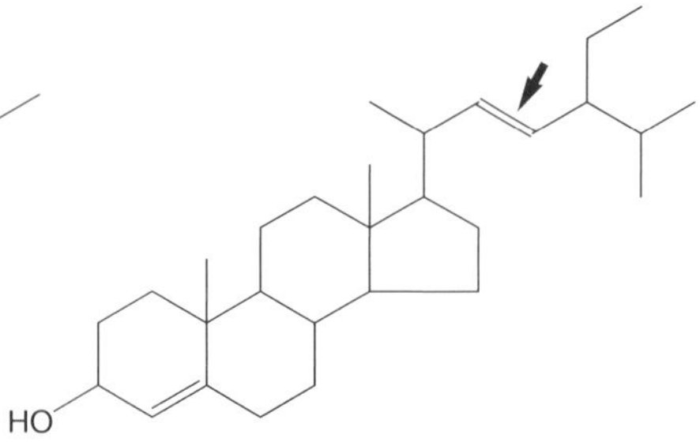

Stigmasterol

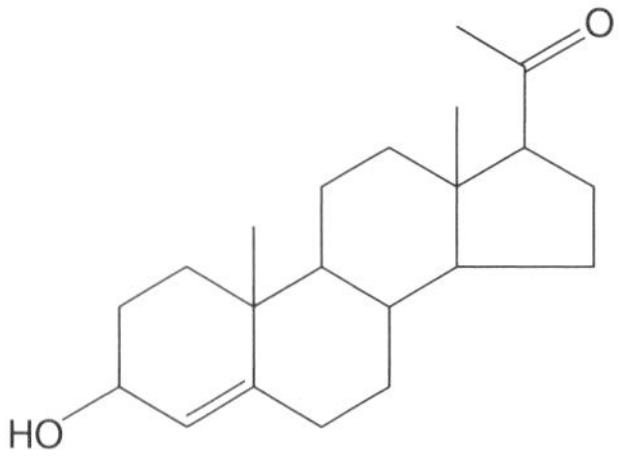

Pregnenolone

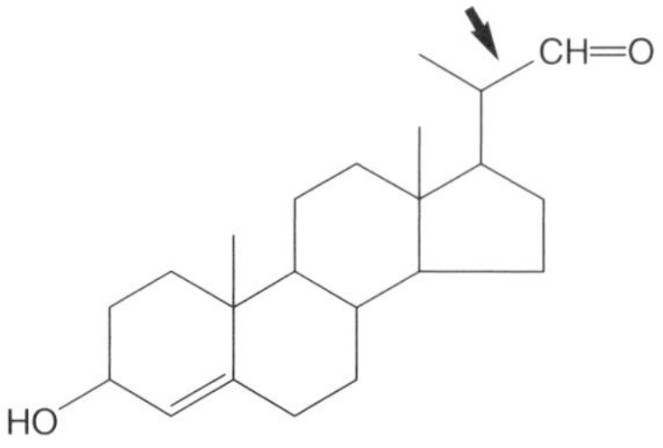

"Bisnoraldehyde"

Figure 25   Sitosterol, stigmasterol, and their conversion to a progesterone intermediate.

left to sit open to the atmosphere in a back lot of the plant in the hope that some microorganism would develop a taste for the sterol and transform it to some useful steroid intermediate. Extensive efforts over the years in many other laboratories did eventually identify a *pseudomonas* microbe that removed the entire side chain to give a mixture of products that included several androstanes with a ketone group at the 17-position.[56]

In the process for converting stigmasterol to progesterone, the alcohol at the 3-position is first oxidized to a ketone in a variant of the 1934 procedure. This protects the ring double bond in the next step. Exposure of this intermediate to pure ozone on a plant scale cleaves the double bond in the side chain. Removal of the single remaining superfluous carbon on the side chain of the "bisnoraldehyde," now in the form of an aldehyde (—CH=O), under special conditions developed by Upjohn chemists, leads in a single step to progesterone.[57,58] The availability of relatively inexpensive progesterone, in concert with the fermentation hydroxylation process and the new route for transforming the side chain at 17, translated into a process for producing cortisone in kilogram quantities.

Cortisone and dihydrocortisone, as well as their 21 acetate derivatives, came into widespread use by the mid-1950s. These compounds were sometimes hailed as miracle drugs as a result of the dramatic relief provided to arthritis patients. Extensive clinical studies showed that these compounds, and the many analogs that were introduced later, had potent activity against inflammation from a variety of causes. The corticoids were also found to have marked anti-allergic effects.

The activity of these compounds interestingly traces back to the arachidonic acid cascade outlined in Chapter 6 (Fig. 9). These steroids intervene at the very first step of the cascade by inhibiting the release of arachidonic acid from fat stores. It has been established that the corticoids inhibit the action of phospholipase A_2, the enzyme that liberates the substance at the very inception of the cascade. The end result of this decrease is reflected in a corresponding diminution of levels of the inflammatory prostaglandin end products. The anti-allergic activity traces back to the arm of the cascade that results in the formation of the leukotrienes that mediate allergic responses. Decrease in arachidonic acid leads to lowered levels of leukotrienes.

Prolonged use of corticoids was, however, found to cause a host of serious side effects, many of which could be traced to the fact that the compounds retained a marked amount of activity on mineral balance. These effects include salt and water retention, loss of potassium, and even elevated blood pressure. Corticoids not unexpectedly suppress the adrenal gland by a feedback mechanism. Sudden cessation of drug would leave patients with seriously depleted levels of essential natural corticoids. Anyone who has taken these drugs will be familiar with the decreasing dosage regimen used at the end of treatment to allow the adrenals to recover their function.

The fast-growing usage of the drugs before these shortcomings were fully appreciated encouraged a large number of laboratories to undertake research aimed at finding their own market entries. When the impact of the side effects was better appreciated, the focus turned to finding corticoids that had a lower ratio of mineralocorticoid to glucocorticoid activity. These programs produced a veritable host of corticoids. The Thirteenth Edition of the Merck Index, for example, lists nonproprietary names for no fewer that 65 compounds in this class. Although some of those agents did have somewhat reduced effects on minerals, none offered the long-sought true spilt between glucocorticoid and mineralocorticoid activity.

One of the first modifications of the cortisone molecule comprised introduction of an additional double bond in the A ring. The product, prednisolone, is some four times more potent than cortisone (Fig. 26). Increases in potency from modification of the structure at one position in the molecule are often additive to changes at some remote location. The additional double bond in ring A is consequently found in almost all of the later corticoids. There is considerable lore in medicinal chemistry circles to the effect that replacement of hydrogen by a fluorine atom in drugs often leads to a more potent derivative. This expectation is borne out in the case of corticoids. Replacement of hydrogen at the 9-position in prednisone by fluorine leads to fluorprednisolone acetate,[59] a drug that shows yet another increase in potency. Inclusion of an additional methyl group at the 6-position in the progesterone molecule led, as noted previously, to an increase in potency. The analogs modification to prednisolone, leads to methylprednisolone,[60] a drug about half again more potent than prednisone. Addition of substituents at the 16-position has also provided several very potent corticoids. Examples include dexamethasone,[61] which has an additional methyl group at position 16. This compound is about 30 times more potent than cortisone. Triamcinolone[62] has a hydroxyl at that position

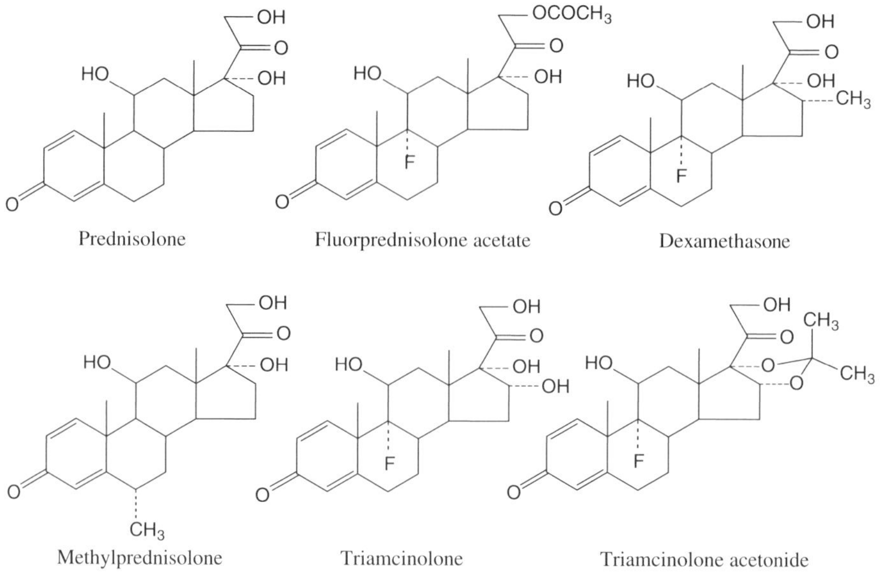

Figure 26   Glucocorticoids.

and shows about five times the potency of cortisone. Simple reaction of that compound with acetone ties up the hydroxyl groups at the 16- and 17-positions in a ring. This derivative, triamcinolone acetonide, is some ten times more potent yet than its immediate precursor. The lower dose required for treatment is a main advantage offered by these drugs. The pattern of side effects is relatively similar across the class.

Glucocorticoids also suppress immune function by a mechanism that is unrelated to phospholipase inhibition. The drugs are, as a result, used to treat patients who have undergone transplants. High levels of drug are usually administered to stave off acute transplant rejection. This property leads also to the use of corticoids in treating autoimmune disorders. The often markedly swollen features of patients receiving the high doses of drug used for those purposes are due largely to fluid retention. This aptly illustrates the remaining mineralocorticoid component of all currently available glucocortioids.

The natural as well as the modified synthetic glucocorticoids retain both the anti-inflammatory and anti-allergic activity when administered topically. Cortisone creams for example are widely used to treat skin rashes due to mild irritation or allergic reactions. Only a vanishingly small amount, if any, is absorbed into the circulation from the site of administration. Those amounts are far too low to exert systemic effects. The fact that cortisone creams are readily available as over-the-counter products attests to the safety of those topical formulations.

Rhinitis is one of the prime markers for allergic reactions. Individuals who suffer from allergies are only too aware of the runny nose and red eyes that greet the coming of spring and the accompanying aerial bombardment with air-borne pollens. The dual properties of corticoids are particularly well suited to treating that condition. These drugs help diminish the initial allergic reaction and, in addition, reverse tissue swelling as a consequence of their anti-inflammatory activity. Several of the more potent corticoids are now available as nasal sprays targeted specifically at allergic rhinitis. Absorption of drugs into the circulation via nasal mucosa is more efficient than through the skin. Despite that, the amount of drug that gets into the bloodstream from these sprays is apparently too low to cause typical systemic steroid effects. One of the first compounds to be formulated as a spray is triamcinolone. Beclomethasone,[63] used in another nasal spray, includes a methyl group at the 16-position that is in the opposite orientation from the corresponding one in dexamethasone; this compound in addition has chlorine at the 9-position instead of fluorine (Fig. 27). The potentiating fluorine atom appears at the 6-position in ring A rather than at 9 in the very potent corticoid flunisolide.[64] Budesonide[65] incorporates a hydroxyl group at the 16-position that is tied up as an acetal with propionaldehyde instead of acetone as in the preceding drug. The potent topical corticoid fluticasone[66] incorporates a modified side chain at the 17-position. The usual hydroymethyl ketone ($-COCH_2OH$) present in most corticoids is replaced in this compound with an ester of a carboxylic acid of a thioalcohol ($-CO_2SCH_2F$). Any amount of this compound that is absorbed in the bloodstream will quickly be hydrolysed to the inactive carboxylic acid.

It is of note that laboratories in the pharmaceutical industry now devote little if any time to the synthesis of new steroids. The very large amount of structure–activity data accumulated over the past decades on the various steroid classes

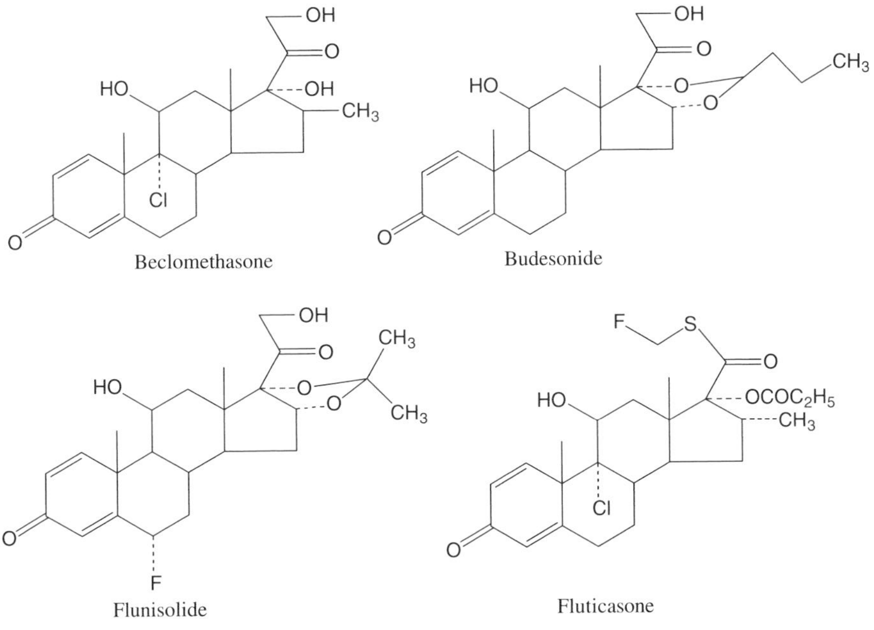

Figure 27   Steroids used in anti-allergy nasal sprays.

have apparently provided a convincing argument that wider splits in activity are not likely to be achieved by further structural manipulations. The argument holds that it is not likely for example that an anabolic agent free of androgenic effects or a corticoid free of mineralocorticoid can be found. Considerable attention is, however, devoted to developing innovative formulations and means of administration.

REFERENCES

1. The historical section draws heavily on L. F. FIESER AND M. FIESER, *Natural Products Related to Phenanthrene*, 3rd ed, Reinhold, NY, 1949.
2. ANON., *FDA Backgrounder on Conjugate Estrogens*, 7/7/1905; available at http://www.fda.gov/cder/news/cebackground.htm.
3. J. W. COOK, E. C. DODDS AND C. I. HEWETT, *Nature*, 131, 56 (1933).
4. M. S. KHARASH AND M. KLEINMAN, *J. Am. Chem. Soc.*, 65, 11 (1939).
5. S. R. SHELTON AND M. G. VANCAMPEN, U.S. patent 2,460,991 (1947).
6. R. E. ALLEN, F. P. PALAPOLI, E. L. SCHUMANN AND M. G. VANCAMPEN, U.S. patent 2,914,563 (1959).
7. G. R. BEDFORD AND D. N. RICHARDSON, *Nature*, 212, 733 (1966).
8. R. J. TOIVOLA, A. J. KARJALAINEN, K. O. A. KURKELA, M.-L. SODERWALL, L.V. M. KANGAS, G. L. BLANCO AND H. K. SUNDQUIST, U.S. patent 4,696,949 (1987).
9. D. LEDNICER, S. C. LYSTER, B. D. ASPERGEN AND G. W. DUNCAN, *J. Med. Chem.*, 9, 172, (1966).
10. C. D. JONES, M. D. JEVNIKAR, A. J. PIKE, L. J. BLACK, A. R. THOMPSON, J. F. FALCONE AND J. A. CLEMENS, *J. Med. Chem*, 27, 1057 (1984).
11. D. GIUDICI, G. ORNATI, G. BRIATICO, F. BUZZETI, P. LOMBARDI AND F. DISALLE, *J. Steroid Biochem.*, 30, 391 (1988).
12. P. N. EDWARDS AND M. S. LARGE, U.S. patent 4,935,437 (1990).
13. ANON., *Drugs Fut.*, 21, 123 (1996).
14. M. LANG, P. BATZL, P. FURET, R. BOWMAN, A. HAUSLER AND A. S. BHATNAGAR, *J. Steroid Biochem. Mol. Biol.*, 44, 421 (1991).
15. J. PROUS, A. GRAUL AND J. CASTANER, *Drugs Fut.*, 19,335 (1994).
16. For a more detailed account of the chemistry and lead references see D. Lednicer, *Strategies for Organic Drug Synthesis and Design*, Wiley, NY, 1998, pp. 86–89.
17. J. C. BABCOCK, E. S. GUTSELL, M. E. HERR, J. A. HOGG, J. C. STUCKI, L. E. BARNES AND W. E. DULIN, *J. Am. Chem. Soc.*, 80, 2924 (1958).
18. D. N. KIRK, V. PETROW AND D. N. WILLIAMSON, *J. Chem. Soc.*, 2821 (1961).
19. H. J. RINGOLD, J. P. RUELAS, E. BATRES AND C. DJERASSI, *J. Am. Chem. Soc.*, 81, 3712 (1959).
20. K. BRUCKNER, B. HAMPEL AND V. JOHNSON, *Chem. Ber.*, 94, 1225 (1961).
21. M. EHRENSTEIN, *J. Org. Chem.*, 9, 435 (1944).
22. L. RUZICKA AND K. HOFFMANN, *Helv. Chim. Acta*, 20, 1280 (1937).
23. A. J. BIRCH AND H. SMITH, *J. Chem. Soc.*, 1882 (1951).
24. F. B. COLTON, U.S. patent 2,655,518 (1952).
25. C. DJERASSI, G. MIRAMONTES, G. ROSENKRANTZ AND F. SONDHEIMER, *J. Am. Chem. Soc.*T, 76, 4092 (1954).
26. M. S. DEWINTER, C. M. SIEGMANN AND C. A. SZPILFOGEL, *Chem. Ind.*, 905 (1959).
27. H. L. DRYDEN, G. M. WEBER AND J. WEICZOVEK, *J. Am. Chem. Soc.*, 86, 742 (1964).
28. G. A. HUGHES AND H. SMITH, *Chem. Ind.*, 1022 (1960).
29. S. N. ANANCHENKO AND I. V. TORGOV, *Dokl. Akad. Nauk. SSR*, 127, 533 (1959).
30. G. H. DOUGLAS, J. M. GRAVES, D. HARTLEY, G. A. HUGHES, B. S. MCLAGHLIN, J. B. SIDELL AND H. SMITH, *J. Chem. Soc.*, 5072 (1963).
31. H. HOFMEISTER, R. WIECHERT, K. ANNEN, H. LAURENT and H. STEIBECK, U.S. patent 4,081,537 (1978).
32. A. V. SCHROFF, U.S. patent, 4,027,019 (1977).

33. K. Ponsold, M. Hubner and M. Oettel, U.S. patent 4,167,517 (1979).
34. L. Veluz, G. Nominee, R. Bucourt and J. Mathieu, *Compt. Rend.*, 257, 569 (1963).
35. For a detailed discussion of the development of this drug see G. Teutsch, R. Deraedt and D. Philbert, in D. Lednicer, Ed., *Chronicles of Drug Discovery*, Vol. 3, ACS Books, Washington, DC, 1993, pp. 1–43.
36. G. Neef, S. Beier, W. Elger, D. Henderson and R. Wiechert, *Steroids*, 44, 349 (1984).
37. G. Neef and R. Wiechert, *J. Steroid Biochem.*, 27, 851 (1987).
38. L. Ruzicka and K. Hoffmann, *Helv. Chim. Acta*, 18, 1487 (1935).
39. C. Meystre, H. Frey, W. Vosser and A. Wettstein, *Helv. Chim. Acta.*, 34, 734 (1956).
40. B. Camerino, M. B. Patelli and G. Salla, U.S. patent 3,060,201 (1962).
41. R. O. Clinton, A. J. Manson, F. W. Stonner, A. L. Beyler, G. O. Potts and A. A. Arnold, *J. Am. Chem. Soc.*, 81, 1513 (1959).
42. R. Pappo and C. J. Jung, *Tetrahedron Lett.*, 365 (1962).
43. M. E. Herr, J. A. Hogg and R. H. Levin, *J. Am. Chem. Soc.*, 78, 501 (1956).
44. A. L. Wilds and N. A. Nelson, *J. Am. Chem. Soc.*, 75, 5366 (1953).
45. F. B. Colton, L. N. Nysted, B. Riegel and A. L. Raymond, *J. Am. Chem. Soc.*, 79, 1123 (1957).
46. G. H. Rasmusson, G. F. Reynolds, N. G. Steinberg, E. Walton, G. F. Patel, T. Liang, M. A. Cascieri, A. H. Cheung, J. R. Brooks and C. Berman, *J. Med. Chem.*, 29, 2298 (1986).
47. I. M. Thompson et al., *New Engl. J. Med.*, 349, 215 (2003).
48. A. Panzeri, L. Ceriani and P. Griggi, U.S. patent 5,342,948 (1994).
49. S. V. Frye, H. N. Bramson, D. J. Hermann, F. W. Lee, A. K. Sinhababu and G. Tian, *Pharm. Biotechnol.*, 11, 393 (1998).
50. L. Fieser and M. Fieser, *Natural Products Related to Phenanthrene*, 3rd edn, Reihold, NY, 1949, p. 406.
51. P. S. Hench, E. C. Kendall, C. H. Slocumb and, H. F. Polley, *Proc. Staff Meet. Mayo Clin.*, 24, 181 (1949).
52. L. H. Sarett, *J. Biol. Chem.*, 162, 601 (1946).
53. D. H. Peterson, H. C. Murray, S. H. Epstein, L. M. Reineke, A. Weintarub and P. D. Meister, *J. Am. Chem. Soc.*, 74, 5933 (1952).
54. J. A. Hogg, P. F. Beal, A. H. Nathan, F. H. Lincoln, W. P. Schneider, B. J. Magerlein, A. R. Hanze and R. W. Jackson, *J. Am. Chem. Soc.*, 77, 4436 (1955).
55. L. Fieser and M. Fieser, *Natural Products Related to Phenanthrene*, 3rd edn, Reihold, NY, 1949, p. 385.
56. M. G. Wovcha, F. J. Antosz, J. C. Knight, L. A. Kominek and T. R. Pyke, *Biochim. Biophys. Acta*, 53, 308 (1978).
57. F. W. Heyl and M. E. Herr, *J. Am. Chem. Soc.*, 72, 2617 (1950).
58. G. Slomp and J. L. Johnson, *J. Am. Chem. Soc.*, 75, 915 (1953).
59. J. Fried, K. Florey, E. F. Sabo, J. E. Herz, A. R. Restivo, A. Borman and F. M. Singer, *J. Am. Chem. Soc.*, 77, 4181 (1955).
60. J. Fried, G. E. Arth and L. H. Sarett, *J. Am. Chem. Soc.*, 81, 1235 (1959).
61. G. E. Arth, P. B. R. Johnston, J. Fried, W. Spooner, D. Hoff, L. H. Sarett, R. H. Silber, H. C. Stoerk and C. A. Winter, *J. Am. Chem. Soc.*, 80, 3161 (1958).
62. S. Bernstein, R. H. Lenhard, W. S. Allen, M. Heller, P. Little, S. M. Stollar, L. I. Feldman and R. H. Blank, *J. Am. Chem. Soc.*, 81, 1689 (1959).
63. J. Elks, P. J. May and N. Galbraith, U.S. patent 3,312,590 (1967).
64. H. J. Ringold and G. Rosenkranz, U.S. patent 3,124,571 (1964).
65. A. Tahlen and R. L. Brattsand, *Arzneim. Forsch.*, 29, 1787 (1979).
66. G. H. Phillipps, B. M. Bain, I. P. Steeples and C. Williamson, U.S. patent 4,335,121 (1982).

Chapter 8

Histamine

Therapeutic categories provided the unifying themes for the initial sections of this volume. The majority of drugs in the preceding chapter shared a common structural feature, the steroid nucleus. In this chapter, a series of drugs, with what appear at first sight quite disparate activities and chemical structures, owed their design and development to a common target. The initial compounds in these various series were aimed at reversing the deleterious effects of histamine (Fig. 1). The role of this structurally very simple compound in allergic reactions had been suggested in 1910 by Dale and Laidlaw at the Wellcome Foundation.[1] Subsequent research on this compound had by the early 1930s confirmed that suggestion. Histamine, which occurs in most tissues, was, for example, found to have profound effects on smooth muscle and blood pressure. Of immediate interest was the finding that the agent is released as the end result of an allergic response. Once in the bloodstream it causes a range of familiar symptoms, from rhinitis to rashes and, in severe cases, anaphylactic reactions.

With the identity of the causative agent of the allergic response confirmed, Bovet at the Pasteur Institute in Paris launched a program to identify compounds that would act as histamine antagonists. Failing to find an endogenous substance that fulfilled that function, he launched a program to synthesize potential antagonists. This research culminated in the discovery of the first safe and effective antihistamine. This compound, pyrilamine,[2] was introduced for human use in 1944. Bovet was awarded the Nobel Prize in 1957 for his many contributions to pharmacology. The chemical structure of that first antihistamine set the pattern for the many other drugs in this class that were to follow. All the first-generation drugs included two six-membered rings, one or both of which is benzene and the other of which might be a pyridine ring and then a basic amine at the end of a two-carbon chain. It was later established that these drugs acted by competing with histamine for its receptor sites. This in effect blocks the various reactions caused by histamine.

The ready acceptance of this drug caused many other firms to start their own programs to find antihistamines. The relatively simple methods required to prepare

New Drug Discovery and Development by Daniel Lednicer
Copyright © 2007 John Wiley & Sons, Inc.

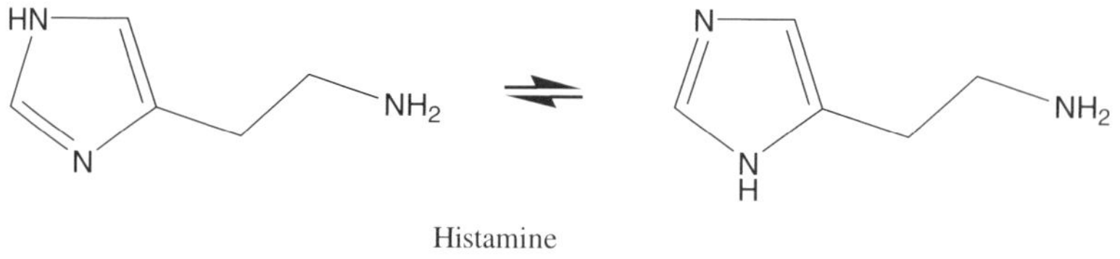

Histamine

Figure 1   Histamine (Note; the compound rapidly equilibrates between two forms, tautomers).

these compounds led to the synthesis of hundreds of analogs. Many of these were introduced into the market at one time or another. Several of those compounds are still in widespread use today (Fig. 2). Chlorpheniramine[3] and doxylamine,[4] for example, make up the antihistamine component of many cold remedies. The antihistamine diphenhydramine[5] is better known as its salt with chlorotheophyline; that combination, Dramamine®, is the motion sickness remedy familiar to those who try to stave off mal-de-mer. The safety and good tolerance of this class of drugs has led to their wide availability over the counter. These compounds are not, however, totally free of side effects. Higher doses of these drugs tend to induce sleepiness.

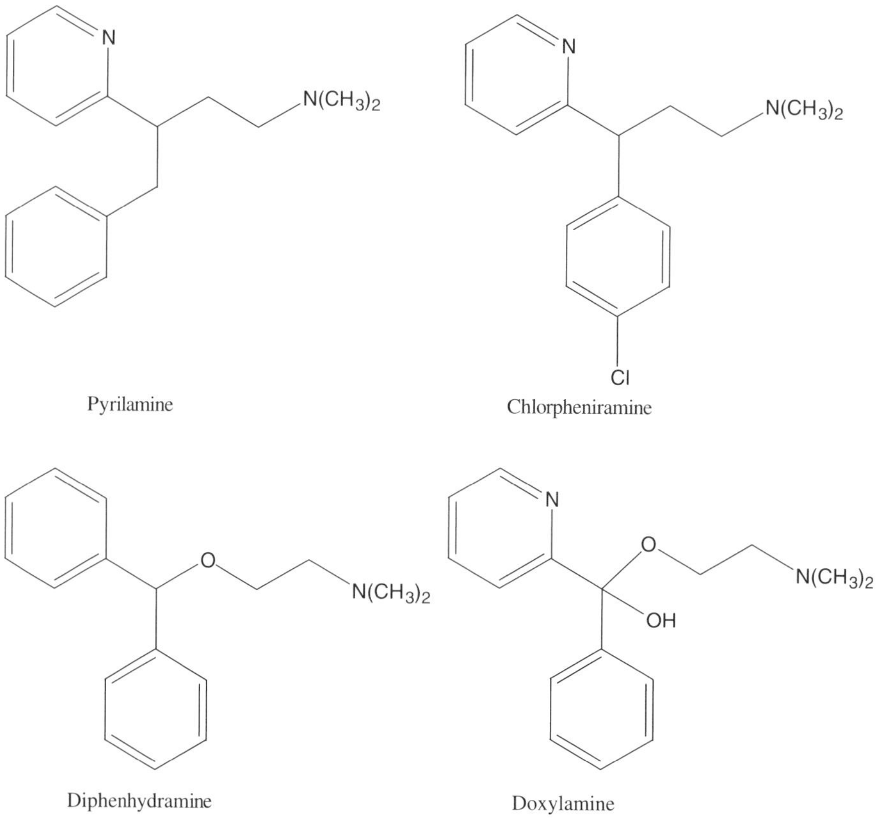

Pyrilamine

Chlorpheniramine

Diphenhydramine

Doxylamine

Figure 2   Open-chain antihistamines.

The active ingredient in one of the popular over-the-counter sleep remedies in fact consists of diphenhydramine.

The notion of preparing rigid analogs from open-chain compounds, which has been noted before, may well have originated with the antihistamines. Within a short time after the introduction of pyrilamine, chemists at Rhone–Poulenc prepared a close analog of the open-chain antihistamines in which two of the benzene rings had been tied back by sulfur into a new fused system called a phenothiazine. The resulting product, diethazine (Fig. 3),[6] surprisingly seems to have lost much much of its antihistamine activity; it has found some use for treating Parkinson's disease, but has been largely replaced by the more effective drugs that are now available. Simply adding a methyl group to the side chain restores antihistamine activity. The compound, promethazine, is still in use and better known by its trade name Phenergan®.

The additional carbon atom in the side chain of promethazine did, however, seem to increase the side effects that could be traced to activity on the central nervous system (CNS). Those effects were even more prominent when the three-carbon-atom side chain was straightened out and the basic nitrogen moved to the end. Animal tests indicated that the CNS activity of this compound, promazine (Fig. 4),[7] had markedly increased in comparison to its antihistaminic activity. Including a chlorine atom on one of the benzene rings increased that ratio even more. That compound, chlorpromazine,[8] introduced the modern treatment of mental illness with psychoactive drugs. The true potential of this drug was for a time obscured by lack of valid animal models for psychoses. According to one account,[9] a French surgeon, Laborit, who was trying to develop a form of anesthesia that he called "hibernation," tried the compound on several patients. He recommended that the drug be investigated as a CNS agent after noting the strange effects of this drug that he likened to "chemical lobotomy." An apocryphal story has it that, arguably prior to this, Rhone–Poulenc offered a license to this drug to a number of companies in the United States. Most firms failed to express any interest in this compound arguably due to its seemingly bizarre activity. Smith Kline French did express interest and took the drug on for further development. Some very perceptive pharmacologists at that firm then performed the proper experiments to deduce the true value of the

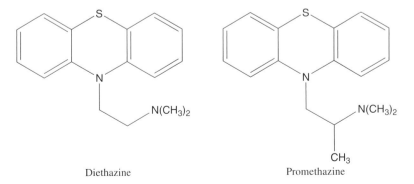

Diethazine   Promethazine

Figure 3   Diethazine and promethazine.

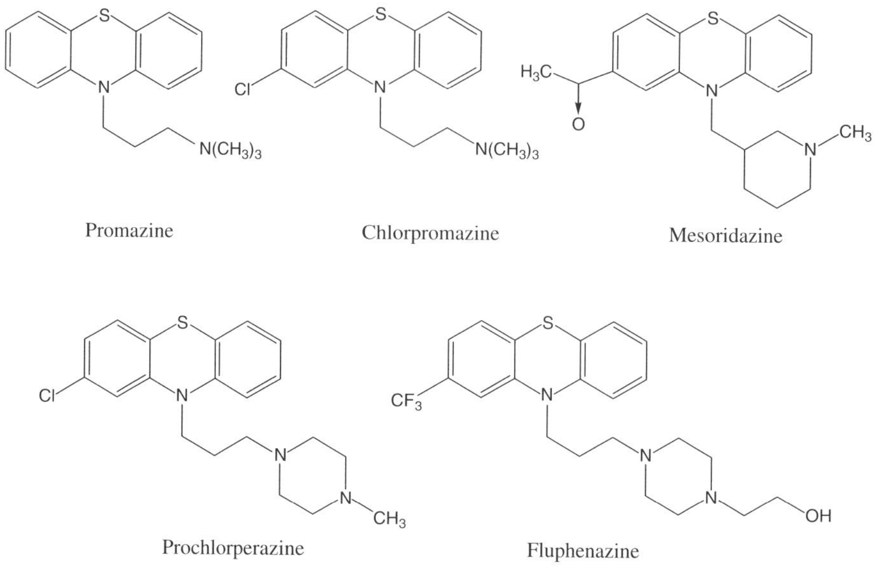

Figure 4   Phenothiazine antipsychotic agents.

drug. Results from those experiments led to its being investigated in patients as an antipsychotic agent. Those trials revealed that this phenothiazine greatly diminished the patients' psychotic episodes. This activity led to these compounds being initially classed as "major tranquilizers." The revolution wrought in the treatment of mental disease by chlorpromazine and the many drugs that came along after has been adequately told many times in other accounts. Promazine itself was revived after the recognition of the activity of its younger congener. That drug too is still available as a low-potency antipsychotic agent.

The discovery of chlorpromazine triggered major programs in competing firms to probe the structural requirements for antipsychotic activity. The number of new chemical entities prepared in those projects probably number in the hundreds. The Merck Index, Thirteenth Edition, lists nonproprietary names for no fewer that 26 phenothiazines. That a reasonable number of these were accepted in clinical practice can be inferred from the fact that at least eight drugs in this class carry FDA approval. The structural requirement for activity in this class allow for some interesting modification. The chlorine atom on the benzene ring in the prototype can be replaced by a number of other groups that withdraw electrons from the ring. The basic nitrogen atom at the end of the apparently obligate three-carbon chain is part of a six-membered piperidine ring in the phenothiazine mesoridazine.[10] Maintaining the three-carbon chain while including that nitrogen in a ring that includes a second basic amine, a piperazine, apparently leads to increased potency. That second nitrogen bears a methyl group as the third substituent in the antipsychotic drug prochlorperazine[11]; this substituent consists of a two-carbon chain that terminates in a hydroxyl group in the drug fluphenazine.[12]

There were no good animal models for psychoses at the time chlorpromazine was first taken into the clinic. The activity of this class of drugs was thus obviously first discovered in humans. This circumstance led to a large amount of effort to develop laboratory models for identifying novel antipsychotic agents and to determine the mechanism by which the phenothiazines achieved their tranquilizing effect. Solution of the first problem led to the development of the large number of new phenothiazines noted above. Information on the mechanism of action of this class of drugs came somewhat later with the development of receptor binding assays. At the risk of oversimplifying, it was found that psychoses are characterized by overabundance of, or oversensitivity to the neurotransmitter dopamine in selected areas of the brain. The phenothiazines were found to have high affinity to dopamine receptors. The drugs competitively bind to those sites and block the neurotransmitter, inhibiting its effects.

The availability of laboratory methods for identifying compounds that had potential antipsychotic activity led to yet another step further from the structure of the first antihistamine compounds (Fig. 5). A group at Jansen Laboratories led by the eponymous scientist undertook a program that involved modification of the nonclassical opiate meperidine (see Chapter 5). One of those analogs, broadly termed a phenyl-piperidine, involved in essence reversing the ester ($-CO_2CH_2 CH_3$) to the alcohol ($-OH$) that would result from hydrolysis. This apparently showed enough activity in a screen for antipsychotic agents to prompt the preparation of additional analogs. One of the compounds from that program, haloperidol,[13] proved to be a major

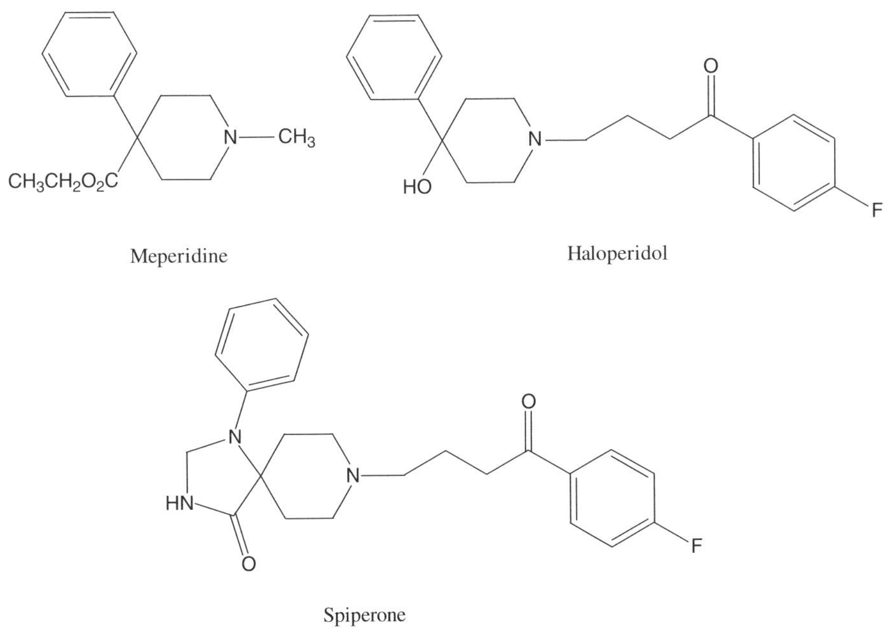

Meperidine   Haloperidol

Spiperone

Figure 5   Phenyl-piperidine antipsychotic agents.

tranquilizer. It achieved very wide use for treating mental illness. This drug was followed by a number of new analogs, some of which reached the clinic. The analog spiperone[14] is one of the most potent known dopamine antagonists and is often used as a tool for labeling dopamine receptors in laboratory studies. It is of interest that drugs in the haloperidol series act by broadly the same mechanism as the phenothiazines. Recent subdivision of the dopamine receptor has, however, pointed to some subtle differences.

The size of a sulfur atom, as noted several times before, is roughly comparable to that of a two-carbon chain. Several instances have been noted earlier in this book where biological activity was retained in molecules that resulted from this interchange (see for example Chapter 6). Chemists at Geigy were gratified to note that the compound that resulted from replacement of sulfur in chlorpromazine by a two-carbon chain, imipramine,[15] retained CNS activity (Fig. 6). This agent was then taken to the clinic on the basis of laboratory data. During those trials, the clinician, Kuhn, noted that the drug caused CNS stimulation in the patients rather than the depression that had come to be expected with phenothiazines. On the basis of that finding he went on to test the drug in depressed patients.[16] Imipramine went on to become the first in a series of drugs for treating clinical depression. These were later known collectively as the tricyclic for the presence in their structure of the three fused rings. The nitrogen atom to which the side chain is attached can be replaced by a carbon atom as in amitriptiline.[17] Further modification, such as replacement of one of the carbons by oxygen in the seven-membered ring gives the antidepressant doxepin.[18] These tricyclic antidepressants also act on dopamine. Rather than blocking the receptors, these agents inhibit re-uptake of the neurotransmitter from the synaptic cleft. The increased levels of dopamine at the receptors tend to lead to mood elevation.

The antidepressant mianserin[19] shares with imipramine the two benzene rings fused to a seven-membered ring (Fig. 7). The basic nitrogen in this, however, is attached to the opposite side of the ring from its position in the typical tricyclics. This agent in fact inhibits re-uptake of serotonin, the other neurotransmitter involved in depression rather than dopamine. This drug, like its tricyclic counterparts, tends to be used largely to treat serious clinical depression. This is due at least in part to the fact that the class as a whole is associated with some often serious side effects. The currently widely used drugs used to elevate mood are quite free of these

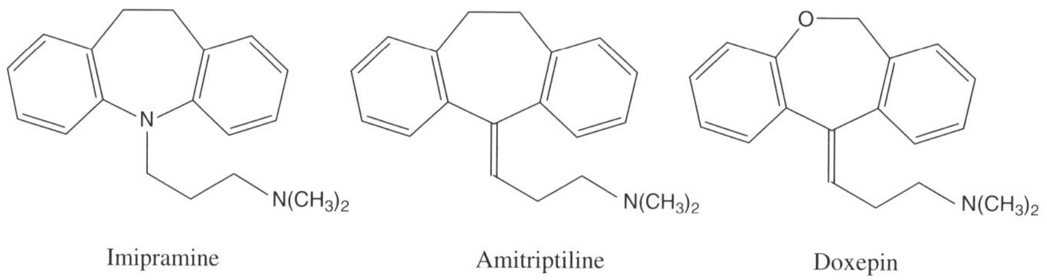

Imipramine   Amitriptiline   Doxepin

Figure 6   Tricyclic antidepressants.

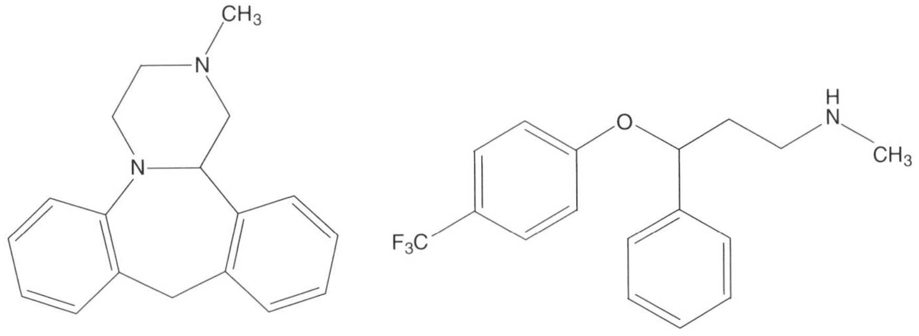

Mianserin   Fluoxetine

Figure 7   Mianserin and fluoxetine.

shortcomings. The structures of these antidepressants, the best known of which is fluoxetine,[20] more familiarly Prozac®, come from a quite different lineage and will not be discussed further. It is of note that the class also acts on serotonin. They are often described as selective serotonin uptake inhibitors, or SSRIs.

Schizophrenia is unfortunately not a passing condition that yields to a finite course of treatment. Control of the disease usually requires that the patient continue to take a maintenance dose of drug for very prolonged periods. A troubling set of side effects emerged as the patient population on long-term maintenance drug increased. These symptoms included tremors, involuntary muscle movements, and, in extreme cases, symptoms characteristic of Parkinson's disease. These so-called extrapyramidal effects are due in part to dopamine deprivation of the neurons in the brain motor system that coordinates movement. A number of more recent antipsychotic agents based on a different tricyclic system have a lower incidence of extrapyramidal side effects. Each of the so-called atypical neuroleptics is based on a central seven-membered ring that includes a nitrogen atom in the two-atom bridge (Fig. 8). These include clozapine,[21] in which the central ring is attached to two benzene rings, and olanzapine,[22] in which one of those rings is replaced by a five-membered ring that incorporates a sulfur atom. The one-atom bridge in the antipsychotic drug quetiapine[23] is replaced by a sulfur atom; the methyl group on the far nitrogen in the piperazine ring is replaced by a long oxygen-containing chain that terminates in an alcohol. The structurally unrelated atypical antipsychotic agent risperidone[24] actually traces its descent from the haloperidol series, although the compound shares only the central six-membered piperidine ring with its ancestor.

The drugs considered thus far came about from a search for agents that would address the involvement of histamine in the symptoms that result from exposure to allergens. The CNS drugs that followed were an outgrowth of side effects of the early anti-allergic antihistamines. It was recognized quite early that gastric acid secretion is among the many physiological responses stimulated by histamine. Stomach acid is, as noted in Chapter 6, intimately involved in the development of gastric ulcers. The fact that the known antihistamines were quite without effect on

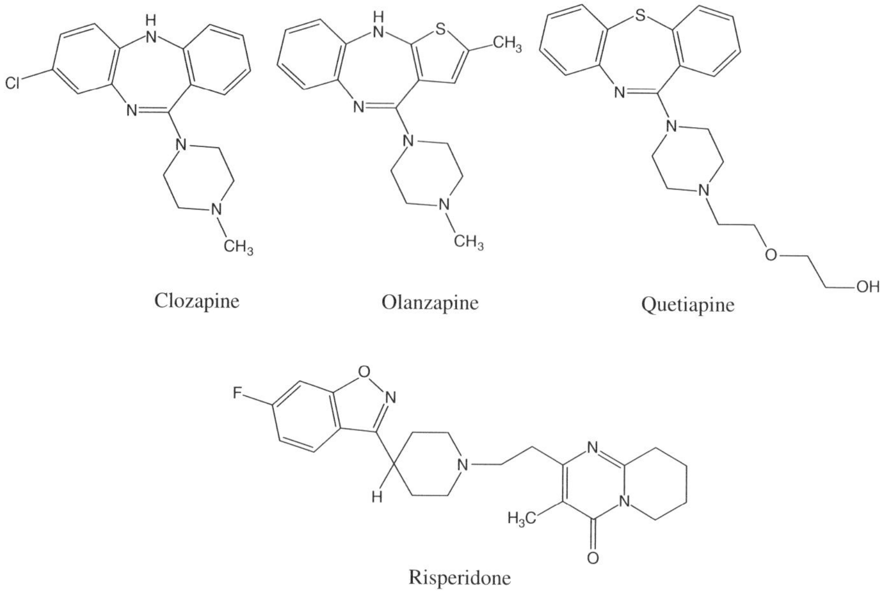

Clozapine   Olanzapine   Quetiapine

Risperidone

Figure 8   Atypical antipsychotic agents.

this end point led to the proposal that the histamine receptor might exist in more than a single form. Gannellin[25] and his colleagues at Smith Kline French in the UK launched a program to synthesize analogs of histamine itself in order to find agents that blocked the putative alternative histamine receptor that controlled stomach acid. James Black, who had also been involved in the work that led to the beta-blocker propranolol, collaborated on this project as well. One of the first compounds that showed at least partial antagonist activity was the derivative of histamine, SK&F 91486 (Fig. 9), in which the carbon chain had been lengthened by one carbon atom; the basic amino group ($-NH_2$) in this analog had been replaced by a guanidine ($NH=CNH(NH_2)$), a group whose basicity is comparable to sodium hydroxide. This agent did, however, retain some agonist activity. Adding yet another carbon atom to the chain and replacing the basic guanidine group by a thiourea ($HNC=S(NH_2)$) finally gave a compound that was a pure antagonist at the histamine receptor that stimulates gastric acid secretion, now dubbed the H_2 receptor. Trials on this agent, burimamide, showed in humans that it was indeed a selective inhibitor of gastric acid secretion. It was, however, very poorly absorbed when administered orally. At this point the chemists added a methyl group to the ring so as to favor the arrangement of double bonds in the five-membered ring responsible for receptor binding; one of the side chain carbon atoms was also replaced by sulfur. The resulting compound metiamide showed good oral activity as an anti-ulcer agent in humans. The drug did, however, cause serious side effects in a small number of patients. Various lines of evidence suggested that this effect was attributable to the presence of the thiourea group. Laboratory studies showed

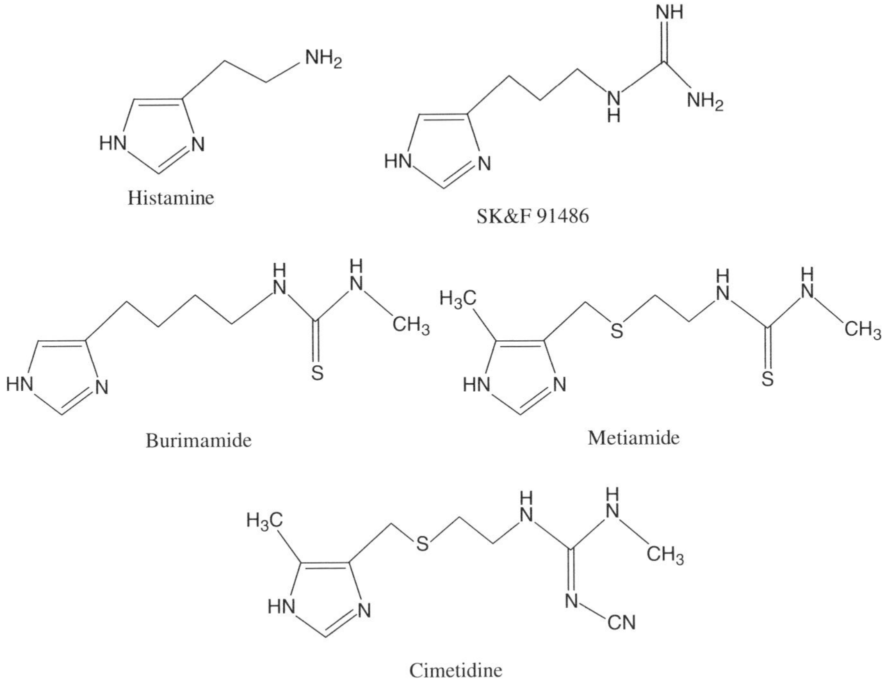

Figure 9   Cimetidine and its predecessors.

that an unusual functional group called a cyanoguanidine had very similar physio-chemical properties to a thiourea. Replacement of this latter group in metiamide by the cyanoguanidine function gave cimetidine, the potent orally active drug that for the first time allowed pharmacological treatment of gastric ulcers (Fig. 9). The drug was also safe enough to find broad use in the treatment of acid reflux. An unver-ifiable account holds that the trade name Tagamet® for cimetidine derives from the fact that the compound was supposedly being groomed as a follow-on successor to metiamide: thus *Tag* along *Met*iamide.

New research programs at both Smith Kline and competitive laboratories were not long in exploring the breadth of the structural requirements for histamine H_2 blockers. Chemists at Bristol Laboratories found that one of the hydrogens on the methyl groups of the cyanoguanidine could be replaced by an acetylene function (Fig. 10). Although this agent, etintidine,[26] is a potent histamine H_2 blocker, it is apparently not used clinically. A more profound change from work at the lab that first produced cimetidine involved replacement of the cyanoguanidine function by a complex heterocycle. It is of note that the aminopyridone at the end of the side chain in oxmetidine[27] incorporates a good part of the guanidine function in cyclized form. In a further demonstration of the looseness of structural requirement, chemists at Allen and Hanbury's were able to replace the five-membered imidazole ring that had descended from histamine itself by a five-membered, oxygen-containing furan ring. The basic nitrogen present in the original ring structure was carried on a carbon

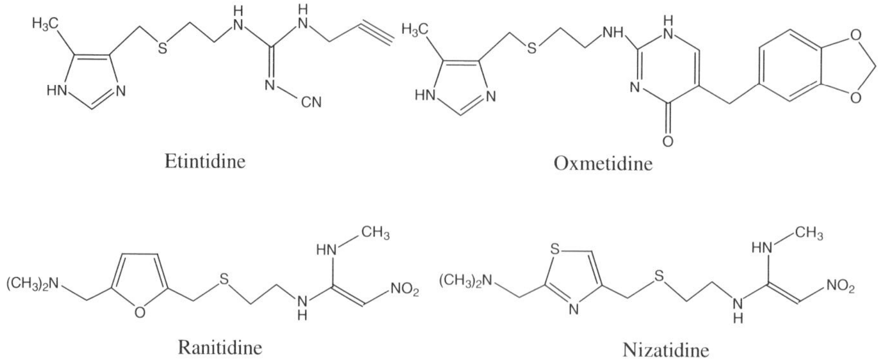

Etintidine   Oxmetidine

Ranitidine   Nizatidine

Figure 10   Second-generation histamine H$_2$ blockers.

atom attached to the ring. The resulting compound also carries a rather different surrogate cyanoguanidine at the end of the side chain. The record of safety of this very widely used drug, ranitidine,[28] has led to it being approved for over-the-counter sales. This approval involves one of the interesting cases where the 75 mg dose is available in almost any store, whereas the 150 mg tablet or capsule is available on prescription basis only. In yet another modification, the furan in ranitidine can be replaced by a thiazole ring, as in the histamine H$_2$ blocker nizatidine.[29]

    The discovery of the histamine H$_2$ blockers stimulated the search for drugs that would inhibit gastric acid secretion by other mechanisms. Not long after cimetidine became available, chemists at Hassle in Sweden launched a program to find compounds that blocked gastric acid secretion by a route that did not involve blockade of the histamine receptor.[30] This effort was encouraged at least in part by shortcomings of cimetidine and its later congeners. Pharmacological tools for the study of gastric acid secretion that had become available as a result of work on the H$_2$ blockers probably helped speed this new research. The project culminated in the discovery of the gastric antisecretory agent omeprazole (Fig. 11).[31] This drug inhibits gastric acid secretion by a mechanism quite distinct from that of the histamine blockers. Stomach acid (that is protons) originates in structures just within the gastric wall called parietal cells. Omeprazole is transformed to a

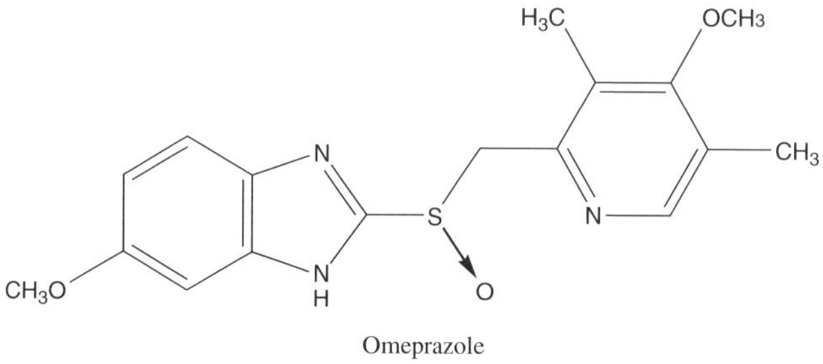

Omeprazole

Figure 11   Omeprazole.

reactive species called a sulfenamide in the relatively acidic space in those cells. This intermediate then interacts with an enzyme system called the proton/potassium (H^+, K^+-ATPase) pump so as to shut down production of acid. The efficacy and safety of omeprazole led to its widespread use. The drug attained approval for over-the-counter sales in almost record time. Drugs from other firms with closely related chemical structures inevitably followed in due time.

The sulfoxide function ($S \rightarrow O$) in the side chain of omeprazole exists in two different forms that bear a mirror image relationship (oxygen above or below the plane of the paper). This then means that the compound itself occurs as two mirror image forms called enantiomers. These are, of course, present in exactly equal amounts in a tablet of omeprazole. Most of the biological activity is in fact due to only one of those forms, as is very often the case with drugs that exist as enantiomers. Hassle, by then part of Astra-Zeneca, went on to develop that single enantiomer, esomeprazole, as a drug in its own right. This patented entity was launched on the market as patents on the parent drug expired.

This chapter opened with the search for drugs that would alleviate allergic reactions by their antihistaminic action. The discussion then traveled afield to drugs developed on the basis of the CNS side effects of some of the early antihistamines. The most prominent of those side effects, sedation, is clearly traceable to a portion of the administered dose that enters the brain. Many an allergy sufferer ended up balancing the misery of severe rhinitis against the likelihood of falling asleep at the wheel of a car. A series of anti-allergic antihistamines that did not enter the brain became available in the mid-1980s. Several different approaches led to these nonsedating antihistamines. They largely rely on the fact that molecules that carry negative or positive charges are far less likely to cross the so-called blood–brain barrier, which regulates entry of compounds into the CNS. Two of these compounds were interestingly found as a result of studies on metabolic transformation of older antihistamines. The rather complex drug terfenadine (Fig. 12) from Richardson Merrell is itself relatively nonsedating. Studies of the metabolism of this drug in humans revealed that it was largely converted to a new compound in which one of the methyl groups was converted to a carboxylic acid ($-CO_2H$).[32] The resulting compound fexofenadine retains full antihistamine activity, but the charge on the acid prevents the drug from crossing the blood–brain barrier. The predecessor, terfenadine, was later withdrawn from the market as it was associated with cardiac toxicity, a side effect not shared by its metabolite. In much the same vein, scientists at UCB Pharmaceuticals in Belgium determined that the venerable antihistamine hydroxyzine was largely metabolized to a derivative in which the alcohol group at the end of the long side chain was oxidized to a carboxylic acid. The resulting compound, cetirizine,[33] is now one of the mainstay nonsedating antihistamines.

An alternative approach to nonsedating compounds comprises incorporating a carboxylic acid or its equivalent into a known antihistamine. Thus, addition of a side chain that terminates in an acid to the old well-known drug triprolidine gives the nonsedating antihistamine acrivastane (Fig. 13).[34] In a similar vein, scientists at Schering–Plough converted the methyl group on one of the nitrogen atoms of the potent antihistamine azatidine to its corresponding carbamate ($-NCO_2CH_2CH_3$).

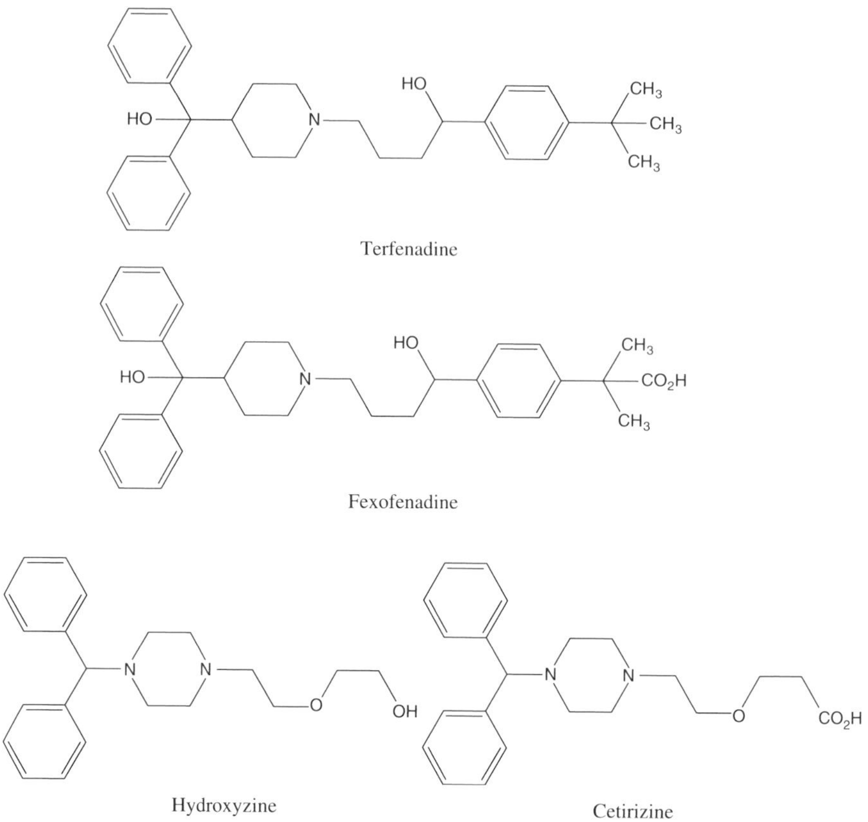

Figure 12 Antihistamine metabolites as nonsedating drugs.

This function is apparently sufficiently polar to keep the drug out of the CNS, even though it does not carry a formal charge. The resulting drug, loratidine,[35] is a widely used nonsedating anti-allergic agent. Some compounds based on rather complicated heterocyclic systems do not cross the blood–brain barrier for other less-defined reasons. The antihistaminic agents astemizole[36] and azelastine[37] have very little sedating properties, in spite of the absence in their structure of groups that carry formal ionic charges.

Attention has more recently turned to the role of leukotrienes in precipitating the symptoms of allergic reactions. These compounds are released by mast cells in response to an allergen in addition to histamine. One of the arms of the arachidonic cascade, as noted in Chapter 6, leads to the formation of the cyclic prostaglandins. The alternative pathway leads to the open-chain leukotrienes. The epoxide (see arrow in Fig. 14) in the first product of this alternative arm of the cascade avidly reacts with various sulfur-containing small peptides. Reaction with glutathione, for example leads to leukotriene C_4. It had long been recognized that a severe allergic reaction proceeded in two phases. The acute immediate phase was

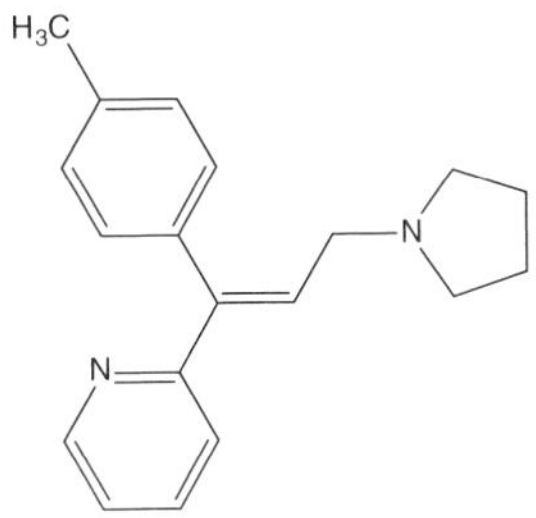

Triprolidine

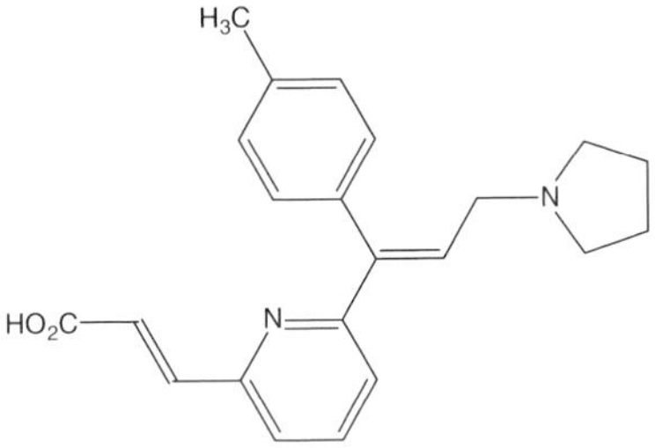

Acrivastane

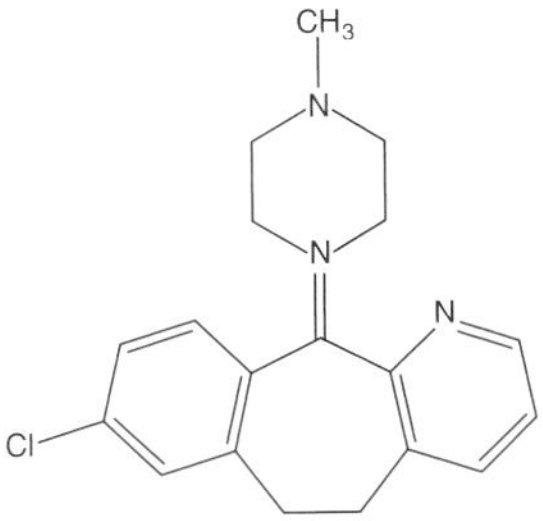

Azatidine

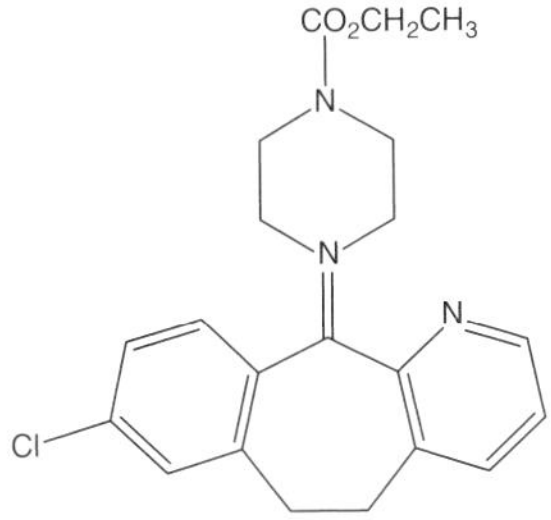

Loratidine

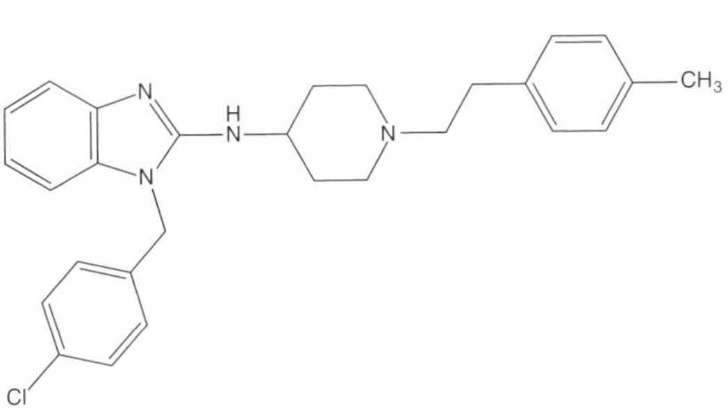

Astemizole

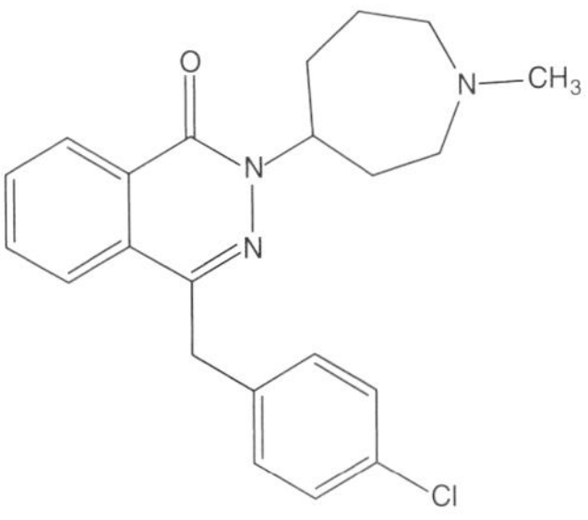

Azelastine

Figure 13 Some more nonsedating antihistamines.

usually traceable to a reaction to histamine. This was followed by the release of an unidentified substance termed slow-reacting substance of anaphylaxis, SRSA. This substance triggered the later events of allergic reactions, including the eponymous anaphylaxis. Progress on the chemistry of products of the arachidonic acid cascade led to the demonstration that SRSA and leukotriene C_4 were one and the same.

Research of leukotrienes has thus far led to regulatory approval of two compounds that act as receptor antagonists. These drugs have proven useful in treating asthma as they have both bronchodilator and anti-inflammatory effects on the

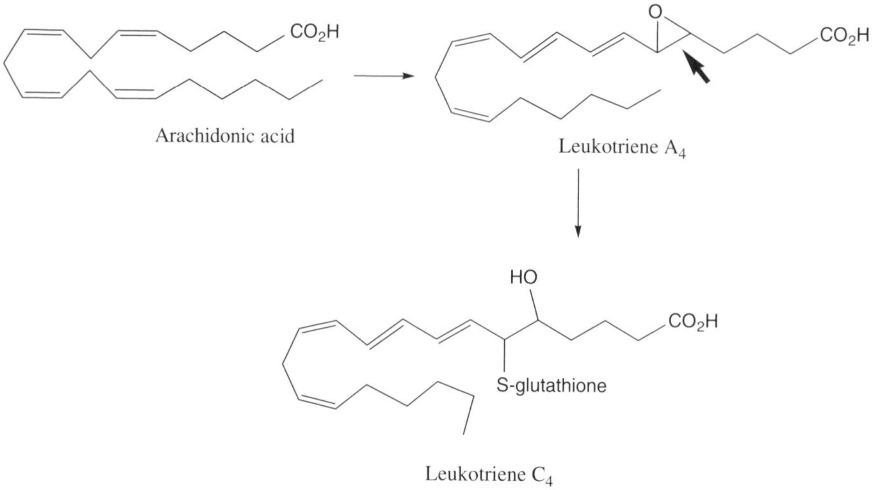

Figure 14   Arachidonic acid and leukotrienes A_4 and C_4.

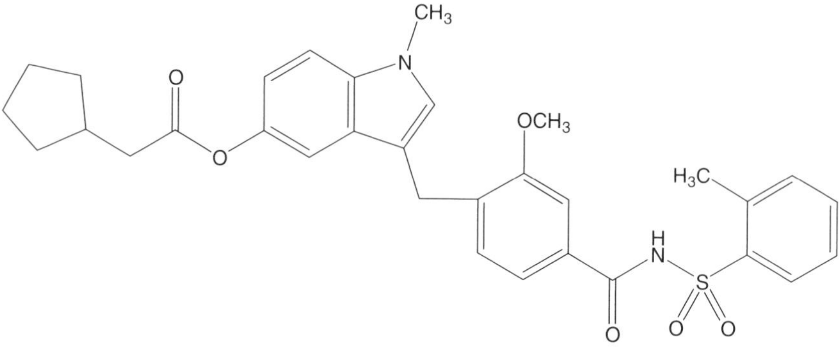

Montelukast

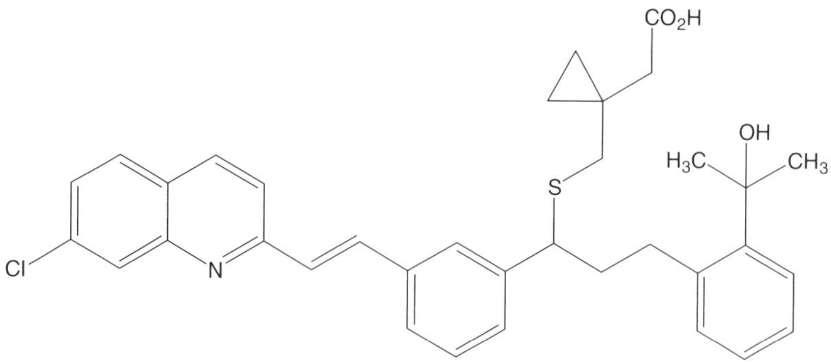

Zafirlukast

Figure 15   Leukotriene receptor antagonists.

airways.[38] The chemical structures of these compounds have little in common beyond the fact that they comprise long extended molecules. The first of these, zafirlukast (Fig. 15),[39] has a relatively polar group at each end, one of which comprises a quite acidic proton (CONHSO$_2$). Montelukast,[40] on the other hand, terminates in a carboxylic acid at one end and a neutral quinoline at the other. The acidic groups in all likelihood mimic the corresponding group present in all leukotrienes.

REFERENCES

1. H. H. DALE AND P. P. LAIDLAW, *J. Physiol. (London)*, 41, 318 (1910).
2. D. BOVET AND A.-M. STAUB, *C. R. Biol.*, 138, 99 (1944).
3. N. SPERBER AND D. PAPA, U.S. patent 2,567,245 (1951).
4. N. SPERBER, D. PAPA, E. SCHWENK AND M. SHERLOCK, *J. Am. Chem. Soc.*, 71, 887 (1949).
5. G. RIEVEASCHL, U.S. patent 2,421,714 (1947).
6. P. CHARPENTIER, *Compt. Rend.*, 225, 306 (1947).
7. P. CHARPENTIER, U.S. patent 2,159,886 (1950).
8. P. CHARPENTIER, P. GAILLIOT, R. JACOB, J. GAUDECHON AND P. BUISSON, *Compt. Rend.*, 235, 59 (1952).
9. H. H. LEHMAN AND T. A. BAN, *Can. J. Psychiatry*, 42, 152 (1997).
10. J. P. BOURQUIN, G. SCHWARB, G. GAMBONI, R. RISCHER, L. RUESCH, U. ULDEMAN, U. THEUSS, E. SCHENKE AND J. REINZ, *Helv. Chim. Acta*, 41, 1072 (1958).
11. R. J. HERCLOIS, U.S. patent 2,902,484 (1959).
12. H. L. YALE AND F. SOWINSKI, *J. Am. Chem. Soc.*, 82, 2039 (1960).
13. P. A. J. JANSSEN, C. vanWESTERHINGE, A. H. M. JAGENAU, P. J. A. DEMOEN, B. F. K. HERMANS, G. H. P. VanDAELE, K. H. L. SCHELLEKENS, A. A. M. vanderEYCKEN and C. J. E. NIEMEGEER, *J. Med. Chem*, 1, 281 (1959).
14. P. A. J. JANSSEN, U.S. patent 3,155,669 (1964).
15. W. SHINDLER AND F. HAFLIGER, *Helv. Chim. Acta*, 37, 472 (1954).
16. R. KUHN, *Schweiz. Med. Wochensch.*, 87, 1135 (1957).
17. R. D. HOFFSOMMER, D. TAUB AND N. L. WENDLER, *J. Org. Chem.*, 27, 4134 (1962).
18. K. STRACH AND F. BICKELHAUPT, *Monatsh.*, 93, 896 (1962).
19. W. J. VAN DER BURG, I. L. BONTA, J. DELOBELLE AND B. VARGRAFTIG, *J. Med. Chem.*, 13, 35 (1970).
20. B. B. MALLOY AND K. K. SCHMIEGEL, U.S. patent 4,314,081 (1975).
21. F. HUNZIKER, E. FISCHER AND J. SCHMUTZ, *Helv. Chim. Acta*, 50, 245 (1967).
22. J. K. CHAKRABARTI, T. M. HOTTEN AND D. E. TUPPER, U.S. patent 5,229,382 (1993).
23. E. J. WARAWA AND B. M. MIGLER, U.S. patent 4,879,288 (1989).
24. L. E. J. KENNIS AND J. VANDENBERK, U.S. patent 4,804,663 (1989).
25. For an informal account of the discovery of cimetidime see C. R. GANNELLIN, in J. S. BINDRA AND D. LEDNICER, Eds, *Chronicles of Drug Discovery*, Vol. 1, Wiley, NY, 1982, pp. 1–38.
26. R. R. CRENSHAW, G. KAVADIAS AND R. F. SANTONGE, U.S. patent 4,157,340 (1979).
27. T. H. BROWN, G. J. DURANT, J. E. EMMETT AND C. R. GANELIN, U.S. patent 4,145,546 (1979).
28. B. J. PRICE, J. W. CLITHEROW AND J. BRADSHAW, U.S. patent 4,128,658 (1978).
29. S. HAYAO, H. J. HAVERA AND W. G. U. STRYKER, U.S. patent 4,006,232 (1977).
30. For a brief history of the discovery of omeprazole see B. A. BERKOWITZ AND G. SACHS, *Molecular Inventions*, 2, 6 (2002).
31. U. K. JUNGGREN AND S. E. SJORSTRAND, U.S. patent 4,255,431 (1981).
32. D. A. GARTEIZ, R. H. HOOK, B. J. WALKER AND R. A. OKERHOLM, *Arznei-Mittelforsch.*, 32, 1185 (1982).
33. E. BALTA, J. DE LANNOY AND L. RODRIGUEZ, U.S. patent 4,525,358 (1985).
34. J. W. A. FINDLAY AND G. G. COKER, U.S. patent 4,501,893 (1985).

35. F. J. VILLANI, C. V. MAGATTI, D. B. VASHI, J. WONG AND T. L. POPPER, *Arznei-Mittelforsch.*, 36, 1311 (1986).
36. F. JANSSENS, R. STOKBROEKX, J. TORREMANS AND M. LUYKX, U.S. patent 4,219,559 (1979).
37. D. VOGELSANG, G. SCHEFLER, N. BROCK AND D. LENKE, U.S. patent 3,813,384 (1974).
38. O. J. DEMPSEY, *Postgrad. Med. J.*, 76, 767 (2000).
39. V. G. MATASA, R. MCEACHERN, H. S. SHAPIRO, B. HESP, D. W. SNYDER, D. AHORNY, R. D. KRELL AND R. A. KEITH, *J. Med. Chem.*, 33, 1781 (1990).
40. M. LABELLE, P. PRASIT, Y. GAREAU, J. Y. GAUTHIER, M. BLOUIN, E. CHAMPION, L. CHARETTE, J. G. DE LUCA ET AL., *Bioorg. Med. Chem. Lett.*, 2, 1141 (1992).

Chapter 9

From Lab Bench to Pharmacy Shelf

A series of complex, interconnected activities will take place between the time that a more or less pure organic compound is identified as a possible drug and the time it shows up on a pharmacist's shelf as an elegant tablet or capsule. Most sponsoring organizations have a set of test criteria for identifying candidates for further development. These will weigh the promise of returns from a new drug against the sizeable expense of bringing such to the market. Once that decision has been made they will in essence first perform additional biological testing to assure themselves that the compound has a good chance of succeeding in the clinic. This will involve more sophisticated models of whatever disease state is being addressed. Considerable attention will also be devoted to elucidating the candidate's mechanism of action down to the molecular biology level if that is possible. Before going much further they will also conduct absorption, disposition, and metabolism studies in animals to gain an understanding as to what happens to the chemical once it is administered. Results from that will guide later formulation, dosing, and toxicology studies. The latter pose a crucial hurdle for any potential new drug, involving both short- and long-term exposure to drug as well as carcinogenicity studies.

When initially identified, the candidate more likely than not comprised of at most a few grams of a powder or some crystals. One of the first steps involves accumulating sufficient amounts to carry through the many other subsequent activities. The compound must also eventually be formulated into a dosage which will ensure absorption and at the same time can be manufactured on today's high-speed machinery. Many of these activities will take place concurrently, precluding a strictly chronological presentation. Once all the required data are at hand, the drug will be taken into the clinic for testing in humans. The compound will then be taken through three carefully designed phases. Close to a decade will have

New Drug Discovery and Development by Daniel Lednicer
Copyright © 2007 John Wiley & Sons, Inc.

passed, on average, between the original discovery and the completion of Phase III trials. If everything has gone well, the data will be submitted to the FDA as a New Drug Application.

This process has undergone many major changes over the years in response to both changing regulatory requirements and the increasing amount of scientific information made available by developments in pharmacology. Back in the 1950s, when many of the compounds detailed in this volume were first introduced, drug sponsors needed to supply only data that addressed drug safety. The Food and Drug Laws then in effect did not require that the sponsor provide data on efficacy. The FDA did, however, indirectly use efficacy as an endpoint by tying safety to the intended use of the candidate drug. The documents of a New Drug Application at that time would as a result often comprise no more than a handful of rather thick binders. Legislation passed in 1962 specifically added the requirement of a demonstration of efficacy for approval. Passage of this law, known as the Kefauver Amendment, was helped in no small way by the concurrent discovery of the severe birth defects caused by the sleep aid thalidomide. Dr Frances Kelsey of the FDA became an instant heroine for delaying approval of that drug in the United States, albeit for the wrong reasons. The efficacy requirement added to the need for additional data that had to be submitted to the agency at both the clinical and pre-clinical level. A series of serious scandals erupted in the early 1970s involving the quality of pre-clinical data submitted to the FDA in support of New Drug Applications (NDA). These ranged from incredibly sloppy data-keeping and recording to outright fraud. Although most of the incidents involved toxicology studies, the FDA instituted a set of requirements that applied to all submitted data. These regulations, called Good Laboratory Practices (GLP), outline detailed procedures for recording and verifying all preclinical laboratory data. It might be noted in passing that other agencies, such as the Environmental Protection Agency, have adopted very similar regulations.

The original assays or screens used by laboratories to identify clinical candidates may or may not be conducted under GLP regulations. However, all data developed after the decision is made to proceed further with a given compound will be conducted under GLP protocols because these data will form part of the NDA package. These additional biochemical and pharmacological tests are all intended to provide the sponsor with the detailed information on the effects and possible side effects of the new drug. It will also help guide later clinical testing. In the case of a novel antibiotic, as a simplistic example, the secondary testing will involve developing a detailed compilation of the activity of the drug on a wide array of microbes, both in vitro and as far as possible in infected animals. Determining the mechanism of action at the molecular pharmacology level will be particularly important in the case of a novel structural type. If the compound is a new analog in a known series, data will be collected to compare the agent with its predecessors. For a drug for controlling circulatory disease, as a further example, detailed studies of the effect on heart functions such as force of contraction, blood flow, and pressure, that is, hemodynamics, will be studied in several animal models. Corresponding models for carrying out advanced pharmacology are available in many other fields. The specifics, which are beyond simple enumeration, will of course depend on the available in-vitro and in-vivo models for the thera-peutic area in question. These studies are intended to elucidate in the best detail possible the exact manner in which the new agent attacks the disease in question as well as its effect on related organ systems. It is also, to some extent, intended to try to avoid unpleasant surprises when the compound is first administered to humans.

The chemistry of the novel compound will usually be reexamined quite early in the development process. Initial samples probably came out of an analog program conducted in an organic chemist's laboratory. The compound in all likelihood met the criteria of purity and identity used for everyday work. The process used to synthesize the compound was also designed to prepare efficiently a large number of analogs rather than the specific compound in question. When it becomes clear that larger amounts will be required, the synthetic route may well be changed so as to reach the target compound in fewer steps and to minimize the use of expens-ive or exotic reagents. The samples of drug substance that will be used to generate all chemical and other preclinical data that will be submitted to the FDA will need to be manufactured according to the Good Manufacturing Practices (GMP). These are a set of regulations for labeling and record-keeping, closely akin to GLP guidelines for the conduct of biological studies. The GMPs also dictate certain aspects of the physical set-up of active drug substance manufacturing facilities. The purity of the larger sample, whether from the new procedure or simply a scaled-up version of the original method, will be investigated in great detail. This usually involves developing analytical techniques such as high-performance liquid chromatography (HPLC). The chemical structures need to be determined for all impurities present at levels higher than 0.1%. The same method can be used to ensure that the total of all impurities in the sample does not exceed 2%. All subsequent batches of drug sub-stances should at least meet the same analytical profile. A set of assays will be devel-oped at this time, by which all subsequent batches will analysed. These assays will then be validated to assure their accuracy, precision, and specificity.

Levels of drug in serum provide the most reliable datum for studying dose–response relations. Many phases of drug development are thus currently accompanied by determination of blood levels. The HPLC methods developed for assaying drug substances can often serve as a starting point for developing methods for measuring drugs in body fluids. These procedures almost invariably start with some extraction procedure to concentrate the drug into a nonaqueous medium. The required sensitivity of course needs to be orders of magnitude lower than that for the drug substance. Detection of the drug is most commonly by single-wavelength ultra-violet (UV) detection. Diode-array detectors have been used to identify peaks when first validating assays. More involved methods that use HPLC coupled to electrospray mass spectrometers have been used for drugs that lack good UV absorbing groups. Capillary gas chromatography (CGC) has been used for a number of drugs because of its very high resolving power coupled with the availability of a number of high-sensitivity detectors that do not depend on the properties of the analyte. This procedure is often used for drugs that lack groups that absorb UV light. The HPLC and CGC methods often feature a limit of quantitation in the nanogram per milliliter range.

Work on the development of a dosage unit of the drug candidate starts reasonably early in the process. The most convenient method for administering drugs from the view of the patient is by mouth. Nothing is simpler than swallowing a tablet or capsule. Many methods for screening drugs in fact involve administering doses to animals by essentially that route. An increasing number of current lead seeking and lead development methods, however, involve in-vitro assays. A crucial step in the development of any compound identified in vitro involves finding whether the drug is available when administered orally. If not, the sponsor is faced with the need for modifying the molecule to make it available from the GI tract, a task that has proven to be no small challenge. An alternative involves delivering the compound as an injectable or by some other route. Strides have recently been made in innovative methods for delivering drugs that are not orally active. It is of note in this regard that several drugs, including the large peptide calcitonin, are now available as intranasal sprays. In the same vein, an inhalable form of insulin was very recently approved. An additional benefit of absorption via nasal mucosa and lungs comes from the fact that this route initially bypasses the liver, the organ in which many drugs are quickly inactivated by so-called first-pass metabolism.

In the case where the new compound is orally active, the sponsor is faced with the choice of formulating the drug as a capsule or as a tablet. These dosage forms are almost universally referred to in the lay press as pills, to the bemusement of pharmaceutical professionals. By their definition, pills comprise a nearly extinct dosage form built up on a seed by adding layer after layer of active drug to the surface while tumbling in a stream of air in a rotating drum called a coating pan.

Capsules generally involve simpler formulations. In the case of low-potency compounds, the fill generally involves pure drug substance milled to some specified particle size. A trace lubricant, also called a glidant, is often added to aid flow of the powder through the high-speed capsule-filling machinery. The drug substance is mixed with some inert diluent such as sorbitol in the case of high-potency drugs that need to be administered in low dose. Disadvantages include the fact the capsule-filling machines are significantly slower than tablet presses and that this formulation involves the added cost of capsule shells.

The salient properties required of a tablet are almost self-contradictory. The final product must be hard and strong enough to withstand rough handling, yet it must disintegrate and promptly release its content on reaching the stomach. In practice, the active drug substance is mixed with a binder such as ethyl cellulose that will hold the tablet together. Additional ingredients may include an inert filler in the case of high-potency drugs and a lubricant to aid flow through the machinery and to help release the tablet from the press. Wetting and disintegration agents are usually added as well to help the tablet come apart once in the liquid environment of the stomach. This complex mixture is then fed into high-speed tablet presses. The powder is subjected in these rotary machines to several tons of pressure between a punch and die at the point where the tablet is formed. A modern 40-station press is capable of turning out a stream of 4400 tablets per minute. An important specification for tablets, as well as capsules, not surprisingly involves maximal time for complete dissolution.

This simply reflects the fact that only that part of a tablet or capsule that is in solution will be absorbed from the GI tract.

Molecules that cannot be administered by mouth or one of the newer innovative routes are usually formulated as injectables. In the case where the compound is stable in water, this may comprise a simple solution sealed in a vial with a special rubber stopper. That bare statement omits the many details involved in developing and validating the required sterile filling protocols. Compounds that are stable in water for extended periods are, however, more the exception than the rule. A large proportion of injectable drugs are thus formulated in freeze-dried, or lyophilized form. In this procedure a solution of the drug is apportioned into vials in the same way as in a sterile fill. Special slotted rubber stoppers are then partly seated in the necks of the vials. These are then set on shelves fitted with cooling channels and the assembly is slid into special hermetically sealable chambers. The fill is then frozen and exposed to a high vacuum. A solid cake of drug remains in the vial as the water departs by sublimation. At the end of the cycle, the chamber is brought back to atmospheric pressure with air or sometimes nitrogen. An ingenious mechanism then pushes the shelves together, in the process seating the stoppers firmly in the necks.

Advances in pharmacology and analytical chemistry have made it possible to study in detail what happens to a drug once it is administered. A formal study of this, known by the acronym ADME, for absorption, distribution, metabolism, and elimination, now forms a regular part of the development of all new drugs. The fact that these studies, of course, initially involve experimental animals presents something of a dilemma; the differences in the way different species handle a given foreign chemical is well documented. Acetaminophen is, for example, lethal to cats at quite low doses. It is, however, usually assumed that the use of two species, most commonly rats and dogs will, if results do not differ drastically, in the end forecast the fate of the drug in humans. Studies of absorption involve determining the level of the compound in blood over the course of time. Those levels may be determined using the analytical techniques noted previously. If those are not yet available, the work may rely on radiolabeled drug. This last is usually indispensable in studying distribution.

Special samples of a drug will be synthesized for this in such a way as to incorporate a radioactive atom, usually carbon (^{14}C) but occasionally (^{3}H) tritium at a chemically stable position. The original route for synthesizing the compound often has to be significantly modified for this purpose. The disposition portion of ADME usually involves studying the concentration of the drug in various tissue compartments after administration. A vivid picture of the location of the drug can emerge from exposing photographic film to microtome thin slices of a small animal that has been fed the radioactive drug. More quantitative studies may identify particular compartments or organs in which the drug tends concentrate. Many agents, for example, tend to be taken up in fatty tissues. This provides long-lasting, and sometimes undesirable, levels of the compound as it slowly leaches out from the fatty tissue stores.

Drug metabolism and subsequent elimination comprises a huge topic in its own right, to which sizeable volumes have been devoted. The following few paragraph thus admittedly barely scratch the surface. The process generally devolves about the fact that most organisms are equipped with elaborate biochemical machinery for disposing of foreign chemicals. The goal of the chemical changes wrought on those chemicals is simply to convert them to a more water-soluble form to ease their expulsion. The products from those changes, the metabolites, are not necessarily less toxic than the starting molecule. In some cases quite the opposite is true. Polynuclear aromatic hydrocarbons (PAHs), as an example, are not per se carcinogenic; the arene oxides produced from those compounds in the liver are potent cancer-causing agents. These chemical transformations sometimes result in the formation of biologically active compounds from inert precursors. This book in fact opened with the unexpected metabolism of the dye prontosil to sulfanilamide, the first specific antibacterial agent. Radiolabeled drug is again of great help in the study of the metabolism of a new drug in that it steers the investigator to the fractions that contain drug-derived compounds. A concerted effort will be devoted by the sponsor to determining the chemical structures of the metabolites. These will often be subsequently synthesized as pure samples so that they can be studied for their biological activity and toxicity. The very large body of information now available on the metabolism of various chemical substances will enable the scientist to make some informed guesses as to the structures of likely metabolites. If the administered drug is an ester, to take a trivial example, the corresponding free carboxylic acid will almost certainly be one of the metabolism products. If the compound, as another example, has a basic amine nitrogen substituted by at least one methyl group, there is a good chance that at least one of the metabolites will be the corresponding amine in which the methyl group has been replaced by hydrogen, in effect removed. The pattern of metabolites will be determined more than a single species in order to determine whether there are any unexpected species-related differences in the way the particular drug is handled. These data will also help select appropriate species for longer term toxicity studies. Excretion studies form a natural part of the drug metabolism studies. These establish the actual amount of the drug and its metabolites that is excreted in urine and found in feces. It also ensures that the larger part of the administered dose is accounted for.

Toxicology presents one of the major hurdles that a potential new drug must clear. Developers will have some rough idea of a new compound's toxicity by the time it is declared a candidate for development. One of the very early tests involves nothing more complex than an LD_{50} test in small rodents. This was simply a roughly calculated dose that was lethal to half the animals. The very small populations, eight to ten animals, used in these assays leads to numbers with very broad confidence intervals. The absolute value is, however, of less interest than the ratio of the dose of drug required to achieve the desired effect, often stated as ED_{50} to the LD_{50}. Even that ratio is not an invariant, as it will vary greatly with the gravity of the condition for which the drug is intended. Therapeutic ratios for drugs used in cancer chemotherapy, for example, tend to be rather small. The corresponding number for a drug that restores hair growth should be correspondingly huge. Once a compound is identified as a potential clinical candidate, the simple LD_{50} will almost certainly be followed by longer term administration of drug to two or more animal species at ascending doses. These tests are intended to more fully explore the toxicity of the compound at sublethal levels as well as to determine its possible deleterious effects on various organ systems. Special attention will almost certainly be focused on those organs on which the drug is known to act. The effects of the drug on the cardiovascular system will be studied up to almost lethal doses, for example, on a compound intended to suppress abnormal heart rhythm. Some classes of drugs have met their doom at this point; a series of very potent progestins were withdrawn because they caused uterine nodules in beagles, an effect subsequently shown to be unique to this animal. These advanced toxicity studies also examine in detail animals that succumbed at high dose to determine the cause of death.

The attrition rate of clinical candidates at this last stage is understandably quite high as the next step involves testing the compound in humans. To take that step, the sponsoring organization files for permission with the FDA. Formal submission of this application is currently often preceded with a meeting with the agency in order to preview the adequacy of the information that will be submitted. The formal document, called an IND for Investigational New Drug, comprises a massive compilation of all the information developed in the laboratory to date. The Chemistry and Manufacturing Section, for example, will contain among other data a detailed description of each of the steps in the synthesis as well as a proof of structure of the active drug substance. The procedure used in preparing the final dosage will include copies of the manufacturing control documents generated in the many steps that lead to the final doses. The IND will also include analytical data on the active drug substance and final dosage forms as well as documents that describe the tests used to validate those procedures. Documents and copies of instrument recordings generated in the multitude of pharmacological studies will also be included in the fat submission. It is of interest as an aside that GLP and GMP regulations were first promulgated in the late 1970s. Investigators were bemused by the requirement that the sponsor submit the "original" copies of all reports, because the regulation came along at just the time the world was switching to computer-driven printers.

The FDA personnel then consider the package to determine whether the data at hand indicate that the new compound can be safely administered to humans. The contemplated Phase I studies are aimed largely at defining safe doses and gathering ADME information in humans. The studies typically involve no more than 20 to 80 individuals. The studies on most drugs are carried out in normal volunteers. In a typical study, those individuals will receive escalating doses of the test compound; the starting dose will be a small fraction of that which would be the therapeutic dose based on animal data. Various body fluids would be collected at quite frequent intervals. This small number of subjects and the fact that they are usually healthy precludes any consideration of efficacy. Antitumor chemotherapy agents provide an exception to this generalization. These agents are usually tested in cancer patients because these compounds are often quite toxic and may in addition be carcinogens. Results from those studies as a result sometimes give early indications of activity. Once the FDA has granted clearance, the test facility assembles its Institutional

Review Board (IRB), a body composed of staff from the facility where the study will be staged, as well as a mandatory outside member, in order to study the detailed test protocol. This document outlines the proposed project in fine detail. The Informed Consent form signed by subjects is an important document in any study. This form usually leans over backwards to outline the various hazards associated with the study and attests to the subject's or patient's free-will participation. This is a far cry from practice as late as the early 1970s. Industry in those years routinely carried out Phase I studies using prisoners. Parke Davis and Upjohn in fact maintained a small free-standing joint Phase I clinic within the walls of the Jackson Correctional facility in Michigan. All such prison testing ended as a result of a number of court rulings that there could be no such thing as free-will participation among the imprisoned, because they were recompensed either monetarily, with time off their sentence, or other rewards. The conduct of clinical trials too is governed by a set of FDA regulations. This set of rules, known by the acronym GCP, set out detailed guidelines for all the steps involved in a clinical trial, from the content of the Informed Consent form to the format of the documentation that will be submitted to the agency at the completion of each phase.

In the ideal case, data from the Phase I trial will show that the candidate compound is well absorbed and almost entirely excreted in a reasonable amount of time. Observed blood levels in humans are in the range that produced the desired pharmacologic effect in test animals. Test subjects, it will be further assumed, did not show any adverse effects until they had received a dose that was many multiples of the projected therapeutic dose. These data inform the sponsor that the drug is safe; it is, however, still mute as to the compound's efficacy. Cancer chemotherapy agents are an exception to this as clinicians will be looking closely for tumor regression. At this point the sponsor will submit the data to the FDA and schedule a Phase I to Phase II meeting.

Phase II clinical trials are intended specifically as probes for therapeutic efficacy. This is the first time that the new compound will be administered to individuals who will potential profit from treatment, that is, patients. These studies will of course be carried out under GCP regulations, including IRB approval and patient informed consent. The design of the trials will vary considerably with the nature of the illness that is to be treated. The double blind experimental design comprises the *sin qua non* for scientific validity. In this paradigm neither the patient nor the clinician know the identity of the dosage form that is being administered. Some of the elaborate labels designed for those trials include a special tear-off or rub strip that allows the investigator to identify the dose in the case of an emergency. Specifying this design still leaves open a question that can be fraught with ethical questions. The design can call for either placebo or some existing drug for treating the illness in question. The choice is quite obvious when dealing with serious bacterial infections. Depriving patients of effective drugs is hardly defensible. The same applies to illness for which there is no effective drug therapy. Use of placebo is unavoidable in those cases. There are, however, a number of gray areas in which the answer is not quite that clear cut. Phase II trials typically involve between one- and three-hundred patients. The study venue will vary with the therapeutic target. Phase II trial of a new minor tranquilizer agent might well be carried out on an outpatient basis by a central clinic. Investigators usual prefer conducting these trials at a single site, as this makes it easier to control variables. This is not always possible in the example of rare diseases. Test of cancer chemotherapy drugs thus more often than not involves patients located in several different clinics. Trial duration, except in the case of drugs intended for long-term chronic use, tend to be on the short side. Results from the trials will then be subject to rigorous statistical analysis. If the activity is confirmed, the drug is formally entered into the next phase of testing.

This next level, Phase III, is in many ways Phase II writ large. The testing is designed to replicate the expected use of the drug in actual clinical practice. Depending again on the nature of the disease, a Phase IIII trial may involve anywhere from several hundred to well over a thousand patients. Because the efficacy will already have been established in the previous phase, these large-scale studies will not necessarily use a controlled double blind design. These studies will, by their very nature, be carried out at multiple sites. The larger population exposed to the drug will also provide data on some of the more common side effects. The GCP regulations apply at this phase as well, because it involves a still unapproved drug.

Successful completion of a Phase III trial culminates in the filing of a New Drug Application with the FDA. This filing essentially provides the agency with all the data that have been gathered about the chemistry, manufacturing, pharmacology, and clinical testing of the new drug, including summaries of the individual case report forms. These applications have become so massive that the FDA now allows sponsors to file them in electronic form. The time taken by the agency to review new drugs increased year by year, both because of an increase in the number of applications and the data that were required. Legislation passed in 1992 imposed a review fee on sponsors. Proceeds are intended specifically to allow the agency to add staff so as to expedite the process. Review times have shortened significantly since then. Legislation in 1997 allowed the FDA to designate selected drugs as important therapeutic advances. Those agents then receive formal Fast Track Review. Whatever the track, the months after filing will be a busy period for the sponsor in answering inevitable questions from the FDA reviewers. When all questions have been satisfactorily addressed, the agency will present the data on the new drug to its Advisory Committee in the relevant therapeutic area. This group of outside experts will then consider the data at an open public meeting and vote on whether the drug should be approved or whether they consider more data are needed. The FDA almost, but not invariably, follow the Committee recommendation.

Receipt of an "approvable" letter from the FDA often comprises the final step in this long process. This letter not infrequently requests some changes in the labeling or in the package insert. This last document is hardly read by anyone and is routinely removed from packages by the pharmacist. It actually consists of a relatively concise summary of the NDA, whose wording is negotiated between the sponsor and the agency.

A very large number of important new drugs in various therapeutic classes were introduced in clinical practice in the early to mid-1960s. These entities became an attractive target to manufacturers of generic drugs in the 1980s as the patent term on these expired. A close reading of the then existing regulations came close to mandating that the generic file perform all the work with those drugs required to file an NDA, including the three phases of clinical trials. Recognizing the waste of resources involved in this process, the agency proposed an interim rule called the "paper NDA." Under these terms, the generic sponsor could replace the clinical part of the NDA with an extensive literature review of clinical publications on the drug in question. This procedure was streamlined in 1984 with the passage of the Waxman–Hatch Act. This replaced the "paper NDA" with the Abbreviated New Drug Application (ANDA). An ANDA in essence comprises two parts. In the application the sponsor supplies a Chemistry and Manufacturing segment that is fully equivalent to that in an NDA. The applicant must demonstrate that the proposed active drug substance is virtually identical to that in the pioneer drug. It is particularly important to demonstrate that it contains no new impurities when compared to the original. This section must also include data to show that the physical chemical properties of the new dosage pass the same acceptance criteria as the original. In practice this applies mostly to dissolution time testing for tablets or capsules. The ANDA must also contain data that demonstrate that the generic drug is bioequivalent to the original. This last usually involves data from a cross-over clinical trial in about 24 to 30 healthy subjects. The blood level profile is measured in each individual after taking the generic drug and the pioneer drug. The various parameters, such as maximum level, time to that level, and decay of drug levels, must be statistically the same. The FDA currently maintains a very large list on the status of drugs whose U.S. patents have or are about to expire. This "Orange Book" is available both online and as hard copy.

BIBLIOGRAPHY

Much of the material presented in this chapter is available in greater detail in any one of a number of monographs. Some of the more recent volumes are listed below.

Analytical Methods

S. Akuja and S. Scipinski, *Handbook of Modern Pharmaceutical Analyis*, Academic Press, San Diego, CA, 2001.
M. Schwartz, *Analytical Method Development and Validation*, Marcel Dekker, New York, NY, 1996.

Methods for Assaying Drugs in Body Fluids

J. Chamberlain, *Analysis of Drugs in Body Fluids*, CRC Press, Boca Raton, FL, 1995.

Dosage Form Development

M. Gibson, *Pharmaceutical Formulation and Preformulation*, CRC Press, Boca Raton, FL, 1998.

ADME

Y. Kwon, *Handbook of Essential Pharmacokinetics, Pharmacodynamics and Drug Metabolism*, Kluwer Academic/Plenum, New York, NY, 2002.

Drug Metabolism

G. G. Gibson, P. Skett, and N. Thorn, *Introduction to Drug Metabolism*, Nelson Thornes; Cheltenham, UK, 2001.
T. Woolf and W. Woolf, *Handbook of Drug Metabolism*, Marcel Dekker, New York, NY, 1999.

FDA Regulations

The Food and Drug Administration maintains a very detailed website at www.fda.gov. Most of the material on regulations for conducting clinical studies and filing the various documents can, with enough persistence, be dug out from those pages.

Appendix

A Word About the Chemical Structural Diagrams

The foregoing narrative is written in a way that can hopefully be appreciated by simply following the text. The reader cannot, however, help but notice that figures full of chemical structural diagrams occur on most of this book's pages. These are intended to provide more details on many of the explanations of the reasoning that went into the design of various drugs. These structures will most probably be welcomed by readers who are fully up to speed on organic chemistry. If this were a text in organic or medicinal chemistry, the individual compounds would be designated in the figures by numbers; those would then be used in the text when referring to those compounds. This has been deliberately avoided in order to underline the intent that the structures serve as supplements rather than substitutes for text.

The following brief instant review is provided for those readers whose exposure to organic chemistry has receded in the mists of time but who are nonetheless interested in at least occasionally consulting the figures. The structures shown in illustrations are depicted in the highly abstract condensed convention currently favored by chemists. This method of drawing structures of organic compounds in essence shows only the carbon skeleton. Atoms other than carbon are inserted as necessary, as are functional groups. The top line in Figure A1, for example, depicts three views of ethyl alcohol. From left to right (a) shows every single atom, (b) a slightly condensed form that omits bonds, and (c) the current convention as used in this book. The second line repeats this for propyl alcohol. Note that each inflection in a line and each ending represents a carbon atom; the protons on those carbons are implicit.

Going on to a slightly more complex structure, Figure A2 depicts a view of the fully saturated ring hydrocarbon cyclohexane in a view (left) that shows all atoms and the same molecule (right) drawn with the convention used in this

New Drug Discovery and Development by Daniel Lednicer
Copyright © 2007 John Wiley & Sons, Inc.

CH_3CH_2OH

ETHYL ALCOHOL

$CH_3CH_2CH_2OH$

PROPYL ALCOHOL

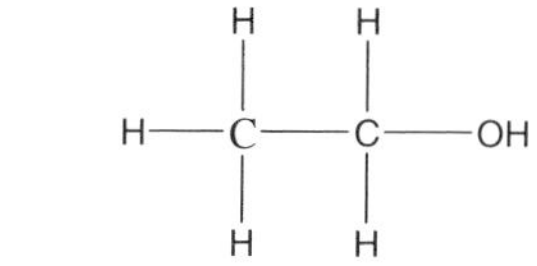

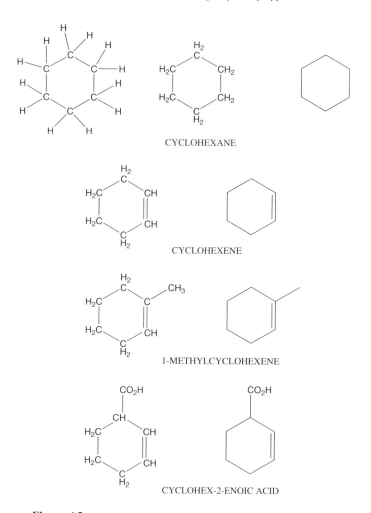

Figure A1 Three ways of showing ethyl and propyl alcohol.

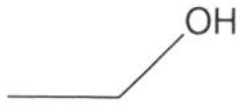

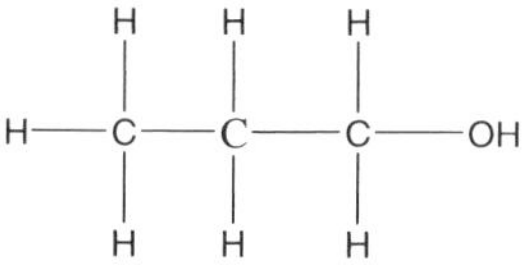

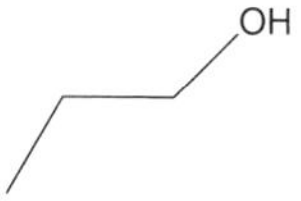

CYCLOHEXANE

CYCLOHEXENE

1-METHYLCYCLOHEXENE

CYCLOHEX-2-ENOIC ACID

Figure A2 Two views of cyclohexane and some of its derivatives.

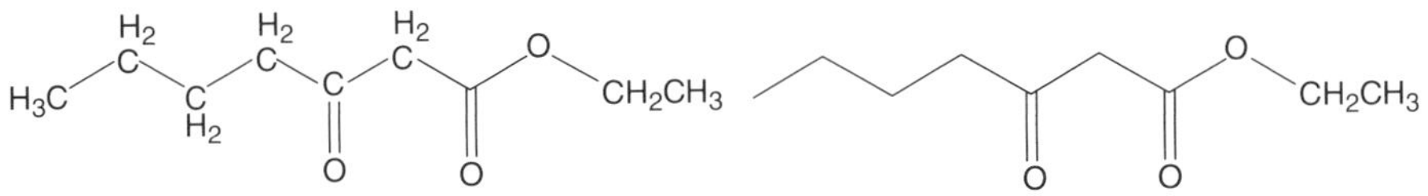

Figure A3   Two ways of looking at an open-chain compound.

book. Corresponding views of several cyclohexane derivatives are depicted on the following lines.

Exactly the same considerations apply to straight-chain compounds (Fig. A3). The skeleton view at right is something of a hybrid. The ester group

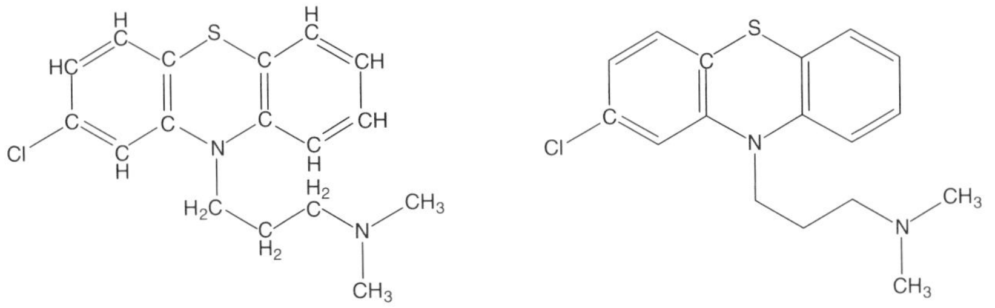

CHLORPROMAZINE

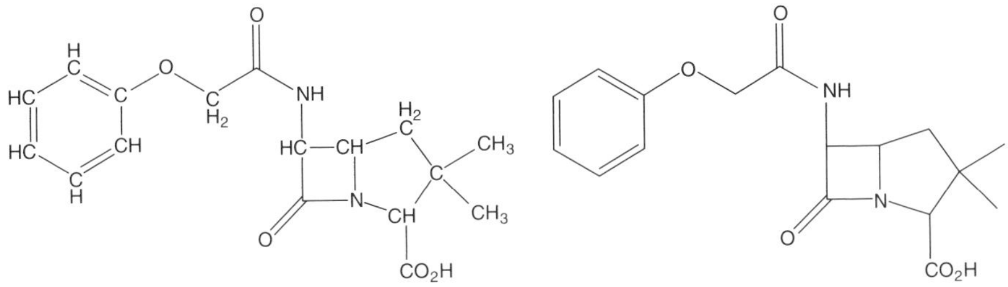

PENICILLIN V

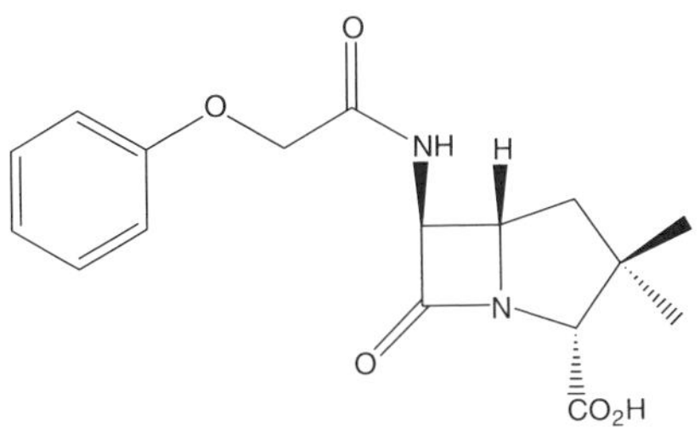

PENICILLIN V 3D

Figure A4   Heteroatoms: chlorpromazine and penicillin V.

($CO_2CH_2CH_3$) is drawn in its fully extended form so as emphasize its presence in the compound.

Compounds that include atoms other than carbon in a ring (hetero atoms) are simply inserted at the proper inflection (Fig. A4).

Organic compounds are of course three-dimensional entities. If this were a book aimed at specialists, the structures of many compounds would include details as to the three-dimensional relationships of various groups. The structure of penicillin V would be depicted as that labeled 3D in Figure A4. These have largely been avoided in order to sidestep yet another level of complexity.

Index

New Drug Discovery and Development by Daniel Lednicer
Copyright © 2007 John Wiley & Sons, Inc.